Digital Ethics Lab Yearbook

Series Editors
Luciano Floridi, Oxford Internet Institute, Digital Ethics Lab,
University of Oxford, Oxford, UK
The Alan Turing Institute, London, UK
Mariarosaria Taddeo, Oxford Internet Institute, Digital Ethics Lab,
University of Oxford, Oxford, UK
The Alan Turing Institute, London, UK

The Digital Ethics Lab Yearbook is an annual publication covering the ethical challenges posed by digital innovation. It provides an overview of the research from the Digital Ethics Lab at the Oxford Internet Institute. Volumes in the series aim to identify the benefits and enhance the positive opportunities of digital innovation as a force for good, and avoid or mitigate its risks and shortcomings. The volumes build on Oxford's world leading expertise in conceptual design, horizon scanning, foresight analysis, and translational research on ethics, governance, and policy making.

More information about this series at http://www.springer.com/series/16214

Carl Öhman • David Watson

Editors

The 2018 Yearbook
of the Digital Ethics Lab

 Springer

Editors
Carl Öhman
Oxford Internet Institute, Digital Ethics Lab
University of Oxford
Oxford, UK

David Watson
Oxford Internet Institute, Digital Ethics Lab
University of Oxford
Oxford, UK

ISSN 2524-7719   ISSN 2524-7727   (electronic)
Digital Ethics Lab Yearbook
ISBN 978-3-030-17151-3   ISBN 978-3-030-17152-0   (eBook)
https://doi.org/10.1007/978-3-030-17152-0

This Springer imprint is published by the registered company Springer Nature Switzerland AG.
The registered company address is: Gewerbestrasse 11, 6330 Cham, Switzerland

Contents

Chapter 1
Digital Ethics: Goals and Approach

Carl Öhman and David Watson

1.1 Digital Technologies As a Force of Good

Are digital technologies a force for good? The question is perhaps somewhat simplistic, but has been posed and rephrased since the early development of computers. As noted by Arthur L. Samuel, one of the great pioneers of Artificial Intelligence (AI), "as to the portents for good or evil which are contained in the use of this truly remarkable machine—most, if not all, of men's inventions are instrumentalities which may be employed by both saints and sinner" (Samuel 1960, p. 742). Indeed, digital technologies have shown a great potential to perpetrate good and evil. From enhancing biomedical research and healthcare (Krutzinna et al. 2018) to improving social interactions (Taddeo and Floridi 2011) and education (Eynon 2013), digital technologies drive major developments of our societies and of individual wellbeing. At the same time, they can enable and exacerbate unfair discrimination (Floridi 2014), undermine fundamental rights (Floridi 2016b; Cath and Floridi 2017; Taub and Fisher 2018), foster mass surveillance (Taddeo 2013), and facilitate cyber warfare (Kello 2017; Taddeo and Floridi 2018) and cyber crime (King et al. 2018). In other words, Samuel was right. As with all technology, the ethical impact of the digital depends on the purposes of its designers and users, the saints and the sinners of Samuel's paper.

Design plays a central role with respect to the ethical impact of technology. Indeed, technological artefacts tend to enable dual uses. However, it is important to stress that, in many cases, artefacts have an "oriented" dual use, which is informed by their design (see Chap. 12 on this concept). They can be used for good or evil,

C. Öhman (✉) · D. Watson
Oxford Internet Institute, Digital Ethics Lab, University of Oxford, Oxford, UK
e-mail: carl.ohman@oii.ox.ac.uk

© Springer Nature Switzerland AG 2019
C. Öhman, D. Watson (eds.), *The 2018 Yearbook of the Digital Ethics Lab*,
Digital Ethics Lab Yearbook, https://doi.org/10.1007/978-3-030-17152-0_1

but seldom can they be used equally well for either. A bayonet may be used for some good, perhaps, but it is really meant to kill a human being. The same holds true for a Swiss army knife: it may have evil applications, but it is designed to provide a set of handy tools. This also applies to digital technologies.

At the same time, the ethical impact of the digital transcends its design and uses. This is because digital technologies transform the reality in which we live by creating a new environment, new forms of (artificial) agency, and new affordances for our interactions with them. Floridi refers to this as to the *cleaving power* of the digital: "the digital 'cuts and pastes' reality, in the sense that it couples, decouples, or recouples features of the world—and therefore our corresponding assumptions about them—which we never thought could be anything but indivisible and unchangeable" (Floridi 2017, p. 123). This the case, for example, of 'real and physical' or 'warfare and violence'. Reality in the information age is no longer coupled to tangibility as much as it is to *interactability* (Taddeo 2012; Floridi 2013). Think of the way in which Alice and her grandfather Bob enjoy their music: Bob may still own a collection of his favourite vinyls, while Alice simply logs into her favourite streaming service (she does not even own the files on her computers). E-books, movies, pictures are all good examples of the decoupling of real and physical in the digital age. Cyber warfare is another compelling case of cleaving power of the digital. For it separates conflict waging from violence (cyber warfare may not cause any casualties or destroy any physical object) (Taddeo and Floridi 2014), agency from responsibility (cyber warfare can be waged by autonomous weapons that are not morally responsible for their actions) (Floridi 2012, 2016c), and undermines state monopoly of political power (grass-roots movements, terrorists, and private companies may all challenge state power in cyberspace) (Nye 2010; Taddeo 2017).

As digital technologies become widely disseminated and their cleaving power reshapes reality and social dynamics, it is crucial to identify the right direction in which to steer this power. To do so, we need to understand what principles should guide the development of current and future information societies, as well as what policies we should enact to ensure that those principles are respected, so that we harness the value of digital innovation to design open, tolerant, equitable, and just information societies. These are ethical questions. Digital ethics is the branch of ethics that addresses them.

Digital ethics "studies and evaluates moral problems related to data (including generation, recording, curation, processing, dissemination, sharing, and use), algorithms (including AI, artificial agents, machine learning, and robots), and corresponding practices (including responsible innovation, programming, hacking, and professional codes), in order to formulate and support morally good solutions (e.g., right conducts or right values)" (Floridi and Taddeo 2016, p. 3).

While they are distinct lines of research, the ethics of data, algorithms, and practices are obviously entangled. One may consider them as three axes defining a conceptual space within which ethical problems may be located. As stressed in Chap. 2, most of the ethical problems posed by digital innovation do not lie on just a single axis. For example, analyses focusing on privacy will also address issues concerning consent and professional responsibilities. Likewise, ethical auditing of algorithms often implies analyses of the responsibilities of their designers, developers, users, and adopters. "Digital ethics addresses the whole conceptual space and hence all three axes together, albeit with different priorities and focus. And for this reason, it needs to be developed as a macroethics, that is, as an overall 'geometry' of the ethical space that avoids narrow, *ad hoc* approaches, but rather addresses the diverse set of ethical implications of digital transformation within a consistent, holistic, and inclusive framework" (Floridi and Taddeo 2016, p. 4).

As a macroethics, digital ethics can provide the necessary guidelines to leverage the transformative power of digital innovation as a force of good. It does so by identifying socially preferable solutions and by assessing trade-offs between conflicting values to inform policy decisions. Digital ethics translates theoretical analyses and principles concerning fundamental ethical issues—like autonomy, human dignity, freedom, tolerance, and justice—into viable guidelines to shape the design and use of digital technologies. The key word here is "translational". Much like translational medicine builds on research advances in biology to develop new therapies and medical procedures, *translational* ethics goes from the whiteboard of academia to the desk of policy making, using theoretical analyses to shape regulatory and governance approaches to digital innovation.

The macroethical and translational approach of digital ethics underpins the vision and work of the Digital Ethics Lab (the DELab). The DELab is a multidisciplinary research environment, which draws on a multitude of academic traditions, including (but not limited to) anthropology, science and technology studies, economics, formal logic, and computer science. Although they may differ in scope and methods, these research areas bring new and important insights that help to identify the ethical problems that arise in our information society and develop the macroethical approach necessary to solve them. At the same time, the DELab has, since the very beginning, established collaborations with a variety of non-academic partners, whose expertise and remits facilitate the translation of ethical analyses into effective guidelines for shaping digital innovation.

1.2   The 2018 Yearbook

The macroethical and translational approach of digital ethics also informs this volume, which collects contributions from the members of the DELab. The goal is to provide an overview of the lines of research undertaken by the DELab and to illustrate the depth and scope of its output. The following 11 chapters of this collection cover a wide range of topics in digital ethics. They highlight the inherently multidisciplinary nature of the subject, which cannot be separated from the epistemological foundations of the technologies themselves or the political implications of the requisite reforms. This is emphasised by Cath, Floridi & Taddeo in the second chapter. The authors provide a helpful overview of the landscape of digital ethics, outlining the major areas of debate, and differentiating the field from related ethical inquiries.

The next three chapters all concern epistemological questions. The third chapter, written by Allo, puts the complex interplay between epistemology and ethics on full display. Allo argues that data science has uncritically imported a collection of so-called "epistemic virtues" from mathematics and statistics, which hinders any meaningful debate around the social structures they inscribe through emerging technologies. In Chap. 4, Turner urges us to take a closer look at how digital technologies evolve in real time, examining the version control platforms Git and GitHub (GitHub 2018). As large scale, collaborative projects are increasingly forged through decentralised networks, Turner argues, researchers in science and technology studies should extend their traditional focus on laboratory studies to online platforms that offer similar, arguably richer resources for investigating the progress of a project. Of course, the largest and best funded scientific projects of our time are major national and international initiatives, like the Large Hadron Collider at CERN. These massive undertakings are the subject of Watson's chapter (Chap. 5), where he draws on classical microeconomic theory to point out potential inefficiencies in contemporary science funding.

How we understand these epistemological and technical questions matters greatly for how we choose to respond to issues that arise from technological innovation. This is illustrated by Chaps. 6, 7, 8, 9 and 10. In Chap. 6, King identifies emerging challenges surrounding AI crime. King analyses the unique and realistic threats posed by AI crime and assesses existing and feasible solutions to mitigate them. Taddeo in turn extends the analysis to cyber-warfare (Chap. 7), arguing that we must "recognise the limits of approaching cyber deterrence by analogy with kinetic conflicts" and move beyond them. In Chap. 8, Cath focuses on the unintended or secondary consequences of technical systems. She points out that the governance and design of such systems are inherently political, advocating for an approach founded on human rights. In pursuing this argument, she identifies a number of biases and gaps in current literature that could be resolved through increasing dialogue and a broader scope.

In Chap. 9, Cowls examines the question of a more specific right, namely that of online privacy. To understand and identify the complex nature of privacy violations, Cowls proposes a framework that applies to three stages of the "information life cycle": collection, analysis, and deployment. He further argues that threats arising at each stage can be considered both on a macro and a micro level, each of which requires its own specific regulatory response strategy. As implied by this chapter, the transition from information society to a *mature* information society (Floridi 2016a) requires massive amounts of new regulation, i.e. legal information. The management and distribution of this information is the topic of Janeček's chapter (Chap. 10). In contrast to the other contributors of the book, Janeček has an explicitly legal, yet also more holistic focus. By first giving a historical overview of the role of ICTs in disseminating legal information, he shows that the information revolution calls for a revision of the current publication and communication model and stresses the importance of designing a system that reaches and communicates well with its ultimate addressees—the citizens who will obey the laws.

Though diverse in scope, these five policy-oriented chapters all illustrate the importance of expert knowledge in the project of designing new reforms and political systems for the digital age. Yet, this task also requires a deep self-understanding in terms of who we are as individuals and as a species. This is the topic of Chaps. 11 and 12.

The information revolution inevitably disrupts some of the most fundamental elements of human existence. For example, Öhman (Chap. 11) argues that the advent of the internet marks a historical shift in how we understand the concepts of life and death. This in turn calls for careful ethical analysis of the macroscopic, microscopic, and conceptual consequences of such a shift. How are we to deal with the informational remains that we leave behind on the web when passing? And what does the prospect of a "digital afterlife" tell us about human existence? Questions of a similar gravity are discussed in the closing chapter, where Floridi presents some "naïve" ideas to help facilitate the formation of a new "human social project" (Chap. 12). He argues that in order to make room for a new, healthier politics, the "Ur philosophy" of the Aristotelian and Newtonian worldview must be abandoned. Instead he proposes that we adopt a "relationist" understanding of the world as the basis of political discourse. In contrast to the rather dark picture painted by many of the chapters, this final chapter shows that there is still room for optimism in digital ethics. The challenges arising from the advent of information society surely seem massive, but so too are human creativity and innovation.

1.3 Conclusion

So, are digital technologies a force for good? Certainly, the cleaving power of the digital offers a wealth of new opportunities and affordances to improve individual wellbeing and foster the development of our societies (Cath et al. 2017). However, these opportunities are coupled with serious ethical risks that can hinder the potential for good of digital technologies. It is crucial to identify and mitigate these risks. The present volume provides the first steps in this direction. Its contributions analyse the opportunities and the ethical challenges posed by digital innovation, delineate new approaches to solve them, and offer concrete guidance to harness the potential for good of digital technologies.

The questions raised here have ancient—perhaps even timeless—roots. While the phenomena they address are inherently new, they are unpacked by examining the fundamental concepts—good and evil, justice and truth—that undergird them all. Indeed, every epoch has its great challenges, and the role of philosophy must be to redefine the meaning of these concepts in light of the particular challenges it faces. This is true also for the digital age. While this book treats important and novel subjects, we know that we have only started redefining and re-implementing fundamental ethical concepts. We look forward to continuing on this journey.

Acknowledgments Before leaving the reader to the chapters in this volume, we wish to thank our colleagues and members of the Digital Ethics Lab at the University of Oxford. Our gratitude goes beyond their contributions to this volume and extends to the many insightful discussions, the cheerful lab meetings, and overall, the enthusiastic and bright research environment in which we can develop our work together. We are also deeply grateful to Springer Nature for publishing the yearbook, and in particular to Ties Nijssen for his interest in the Digital Ethics Lab and support during the process that led to the publication of this volume.

References

Cath, C., and L. Floridi. 2017. The design of the internet's architecture by the Internet Engineering Task Force (IETF) and human rights. *Science and Engineering Ethics* 23 (2): 449–468. https://doi.org/10.1007/s11948-016-9793-y.

Cath, C., S. Wachter, M. Taddeo, and L. Floridi. 2017. Artificial intelligence and the "Good Society": The US, EU, and UK approach. *Science and Engineering Ethics*, March. https://doi.org/10.1007/s11948-017-9901-7.

Eynon, R. 2013. The rise of big data: What does it mean for education, technology, and media research? *Learning, Media and Technology* 38 (3): 237–240. https://doi.org/10.1080/17439884.2013.771783.

Floridi, L. 2012. Distributed morality in an information society. *Science and Engineering Ethics* 19 (3): 727–743. https://doi.org/10.1007/s11948-012-9413-4.

———. 2013. *Ethics of information*. Oxford/New York: Oxford University Press.

———. 2014. Open data, data protection, and group privacy. *Philosophy & Technology* 27 (1): 1–3. https://doi.org/10.1007/s13347-014-0157-8.

———. 2016a. Mature information societies—a matter of expectations. *Philosophy & Technology* 29 (1): 1–4. https://doi.org/10.1007/s13347-016-0214-6.

———. 2016b. On human dignity as a foundation for the right to privacy. *Philosophy & Technology* 29 (4): 307–312. https://doi.org/10.1007/s13347-016-0220-8.

———. 2016c. Faultless responsibility: On the nature and allocation of moral responsibility for distributed moral actions. *Philosophical Transactions of the Royal Society A: Mathematical, Physical and Engineering Sciences* 374 (2083): 20160112. https://doi.org/10.1098/rsta.2016.0112.

———. 2017. Digital's cleaving power and its consequences. *Philosophy & Technology* 30 (2): 123–129. https://doi.org/10.1007/s13347-017-0259-1.

Floridi, L., and M. Taddeo. 2016. What is data ethics? *Philsophical Transaction of the Royal Society A* 20160360: 1–4.

Github. 2018. https://github.com/. Accessed 23 Apr 2019.

Kello, L. 2017. *The virtual weapon and international order*. New Haven: Yale University Press.

King, T., N. Aggarwal, M. Taddeo, and L. Floridi. 2018. *Artificial intelligence crime: An interdisciplinary analysis of foreseeable threats and solutions*, SSRN scholarly paper ID 3183238. Rochester: Social Science Research Network. https://papers.ssrn.com/abstract=3183238.

Krutzinna, J., M. Taddeo, and L. Floridi. 2018. *Enabling posthumous medical data donation: A plea for the ethical utilisation of personal health data*, SSRN scholarly paper ID 3177989. Rochester: Social Science Research Network. https://papers.ssrn.com/abstract=3177989.

Nye, J. 2010. *Cyber power*. Boston: Harvard Kennedy School, Belfer Center for Science and International Affairs. http://www.belfercenter.org/sites/default/files/legacy/files/cyber-power.pdf.

Samuel, A.L. 1960. Some moral and technical consequences of automation—A refutation. *Science* 132 (3429): 741–742. https://doi.org/10.1126/science.132.3429.741.

Taddeo, M. 2012. Information warfare: A philosophical perspective. *Philosophy and Technology* 25 (1): 105–120.

———. 2013. Cyber security and individual rights, striking the right balance. *Philosophy & Technology* 26 (4): 353–356. https://doi.org/10.1007/s13347-013-0140-9.

———. 2017. Cyber conflicts and political power in information societies. *Minds and Machines* 27 (2): 265–268. https://doi.org/10.1007/s11023-017-9436-3.

Taddeo, M., and L. Floridi. 2011. The case for E-trust. *Ethics and Information Technology* 13 (1): 1–3. https://doi.org/10.1007/s10676-010-9263-1.

———. 2014. The ethics of information warfare—An overview. In *The ethics of information warfare*, Law, governance and technology series. New York: Springer.

———. 2018. Regulate artificial intelligence to avert cyber arms race. *Nature* 556 (7701): 296–298. https://doi.org/10.1038/d41586-018-04602-6.

Taub, A., and M. Fisher. 2018. Where countries are tinderboxes and Facebook is a match. *The New York Times*, 21 April 2018, sec. Asia Pacific. https://www.nytimes.com/2018/04/21/world/asia/facebook-sri-lanka-riots.html.

Chapter 2
Digital Ethics: Its Nature and Scope

Luciano Floridi, Corinne Cath, and Mariarosaria Taddeo

2.1 Digital Ethics As a Macroethics

The digital revolution provides huge opportunities to improve private and public life, and our environments, from health care to smart cities and global warming. Unfortunately, such opportunities come with significant ethical challenges. In particular, the extensive use of increasingly more data—often personal, if not sensitive (Big Data)—the growing reliance on algorithms to analyse them in order to shape choices and to make decisions (including machine learning, AI, and robotics), and the gradual reduction of human involvement or oversight over many automatic processes, pose pressing questions about fairness, responsibility, and respect of human rights.

These ethical challenges can be addressed successfully by fostering the development and application of digital innovations, while ensuring the respect of human rights and the values shaping open, pluralistic, and tolerant information societies. Striking such a balance is neither obvious nor simple. On the one hand, overlooking ethical issues may prompt negative impact and social rejection. This was the case, for example, with the NHS care.data programme, a failed project in England to extract data from GP surgeries into a central database. On the other hand, overemphasizing the protection of individual or collective rights in the wrong contexts may lead to regulations that are too rigid, and this may harm the chances to harness the social value of digital innovation. The LIBE amendments, initially proposed to the European Data Protection Regulation, offer a good example, as they would have made data-based medical

L. Floridi (✉) · C. Cath · M. Taddeo
Oxford Internet Institute, Digital Ethics Lab, University of Oxford, Oxford, UK

The Alan Turing Institute, London, UK
e-mail: luciano.floridi@oii.ox.ac.uk

© Springer Nature Switzerland AG 2019
C. Öhman, D. Watson (eds.), *The 2018 Yearbook of the Digital Ethics Lab*,
Digital Ethics Lab Yearbook, https://doi.org/10.1007/978-3-030-17152-0_2

research more difficult.[1] *Social preferability* must be the guiding principle to strike a robust ethical balance for any digital project with impact on human life.

The demanding task of digital ethics is navigating between social rejection and legal prohibition in order to reach solutions that maximise the ethical value of digital innovation to benefit our societies, all of us, and our environments. To achieve this, digital ethics builds on the foundation provided by Computer and Information Ethics, which has focused, for the past 30 years, on the challenges posed by information and communication technologies (Floridi 2013; Bynum 2015). This valuable legacy grafts digital ethics onto the great tradition of ethics more generally. At the same time, digital ethics refines the approach endorsed in Computer and Information Ethics, by changing the Levels of Abstraction (LoA) of ethical enquiries from an information-centric (LoA_I) to a digital-centric one (LoA_D).

The method of abstraction is a common methodology in Computer Science (Hoare 1972) and in Philosophy and Ethics of Information (Floridi 2008, 2011). It specifies the different perspective from which a system can be analysed, by focusing on different aspects, called observables. The choice of the observables depends on the purpose of the analysis and determines the choice of LoA. Any given system can be analysed at different LoAs. For example, an engineer interested in maximising the aerodynamics of a car may focus upon the shape of its parts, their weight, and the materials. A customer interested in the aesthetics of the same car may focus on its colour and the overall look, while disregarding the engineer's observables.

Ethical analyses are developed at a variety of LoAs. The shift from Information (LoA_I) to Digital (LoA_D) is the latest in a series of changes that characterise the evolution of Computer and Information Ethics. Research in this field first endorsed a human-centric LoA (Parker 1968), which addressed the ethical problems posed by the dissemination of computers in terms of professional responsibilities of both their designers and users. The LoA then shifted to a computer-centric one (LoA_C) in the mid 1980s (Moor 1985), and changed again at the beginning of the second millennium to LoA_I (Floridi 2006).

These changes responded to rapid, widespread, and profound technological transformations. And they had important conceptual implications. For example, LoA_C highlighted the nature of computers as universal and malleable tools, making it easier to understand the impact that computers could have on shaping social dynamics and on the design of the environment surrounding us (Moor 1985). Later on, LoA_I shifted the focus from the technological means (the hardware: computers, mobile phones, etc.) to the content (information) that can be created, recorded, processed, and shared through such means. In doing so, LoA_I emphasised the different moral dimensions of information—i.e., information as the source, the result, or the

---

[1] European Parliament, Committee on Civil Liberties, Justice and Home Affairs. (2012). On the proposal for a regulation of the European Parliament and of the Council on the protection of individual with regard to the processing of personal data and on the free movement of such data (General Data Protection Regulation) (COM(2012)0011 – C7-0025/2012-2012/0011(COD)). Amendments 27, 327, 328, and 334–3367 proposed in the Albrecht's Draft Report, Retrieved from http://www.europarl.europa.eu/meetdocs/2009_2014/documents/libe/pr/922/922387/922387en.pdf

target of moral actions—and led to the design of a macro-ethical approach, able to address the whole cycle of information creation, sharing, storage, protection, usage, and possible destruction (Floridi 2006, 2013).

We have come to understand that it is not a specific technology (now including online platforms, cloud computing, Internet of Things, AI, and so forth), but the whole ecosystem created and manipulated by any digital technology that must be the new focus of our ethical strategies. The shift from *information* ethics to *digital* ethics highlights the need to concentrate not only on what is being handled as the true invariant of our concerns but also on the general environment (infosphere), the technologies and sciences involved, the corresponding practices and structures (e.g. in business and governance), and the overall impact of the digital world broadly construed. It is not the hardware that causes ethical problems, it is what the hardware does with the software, the data, the agents, their behaviours, and the relevant environments that prompts new ethical problems. Thus, labels such as "robo-ethics" or "machine ethics" miss the point. We need a digital ethics that provides a holistic approach to the whole universe of moral issues caused by digital innovation.

LoA$_D$ brings into focus the different moral dimensions of the whole spectrum of digital realities. In doing so, it highlights that ethical problems—such as anonymity, privacy, responsibility, transparency, and trust—concern a variety of digital phenomena, and hence they are better understood at that level. So, digital ethics is best understood as the branch of ethics that studies and evaluates moral problems related to *information and data* (including generation, recording, curation, processing, dissemination, sharing, and use), *algorithms* (including AI, artificial agents, machine learning, and robots), and corresponding *practices* and *infrastructures* (including, responsible innovation, programming, hacking, professional codes, and standards), in order to formulate and support morally good solutions (e.g., right conduct or right values). This means that the ethical challenges posed by the digital revolution can be mapped within the conceptual space delineated by three axes of research: the ethics of data/information, the ethics of algorithms, and the ethics of practises and infrastructures.

The ethics of data focuses on ethical problems posed by the collection and analysis of large datasets and on issues ranging from the use of Big Data in biomedical research and social sciences (Mittelstadt and Floridi 2016), to profiling, advertising, data donation, and open data. In this context, key issues concern possible re-identification of individuals through data-mining, −linking, −merging, and re-using of large datasets, and risks for so-called "group privacy", when the identification or profiling of types of individuals, independently of the de-identification of each of them, may lead to serious ethical problems, from group discrimination (e.g. ageism, racism, sexism) to group-targeted forms of violence (Floridi 2014; Taylor et al. 2017). Trust (Taddeo 2010; Taddeo and Floridi 2011) and transparency (Turilli and Floridi 2009) are also crucial topics, in connection with an acknowledged lack of public awareness of the benefits, opportunities, risks, and challenges associated with the digital revolution. For example, transparency is often advocated as one of the measures that may foster trust. However, it is unclear what information should be made transparent and to whom information should be disclosed.

The ethics of algorithms addresses issues posed by the increasing complexity and autonomy of algorithms broadly understood (e.g., including AI and artificial agents such as Internet bots), especially in the case of machine learning applications. Crucial challenges include the moral responsibility and accountability of both designers and data scientists with respect to unforeseen and undesired consequences and missed opportunities (Floridi 2012, 2016); the ethical design and auditing of algorithms; and the assessment of potential undesirable outcomes (e.g., discrimination or the promotion of anti-social content).

Finally, the ethics of practices (including professional ethics and deontology) and infrastructures addresses questions concerning the responsibilities and liabilities of people and organisations in charge of data processes, strategies, and policies, including businesses and data scientists, with the goal of defining an ethical framework to shape professional codes that may ensure ethical practises fostering both the progress of digital innovation and the protection of the rights of individuals and groups (Cath et al. 2017). Here four issues are central: solutions by design, consent, user privacy, and secondary use or repurposing.

While they are distinct lines of research, the ethics of data, algorithms, practices and infrastructures are obviously intertwined, and this is why it may be preferable to speak in terms of three axes defining a conceptual space within which ethical problems are like points identified by three values. Most of them do not lie on a single axis. For these reasons, Digital Ethics must address the whole conceptual space, albeit with varying priorities and foci. As such, it needs to be developed from the start as a *macroethics*, that is, as an overall "geometry" of the ethical space that avoids narrow, *ad hoc* approaches, but rather addresses the diverse set of ethical implications of digital realities within a consistent, holistic, and inclusive framework. An example may help to clarify the value of the holistic approach of Digital Ethics. Consider the case of protecting users privacy on social media platforms. In order to be able to understand the problem adequately and to indicate possible solutions, ethical analyses of users' privacy will have to address data collection, access, and sharing. They will have to focus also on users' consent and the responsibilities on online service providers (Taddeo and Floridi 2015, 2017). At the same time, aspects such as ethical auditing of algorithms and oversight mechanism for algorithmic decisions will be central to the analyses, as they may be part of the solution.

In order to give a better sense of the specific issues discussed in Digital Ethics, the following sections focus on two key areas of application: Internet Infrastructure, and Cyber Conflicts. They are not the only ones, but they should provide a clear sense of the scope and significance of this new area of investigation.

2.2   Digital Infrastructure

Increasingly, digital data infrastructures, like the Internet, are part of what makes our societies prosper (Castells 2007). And as this network-of-networks becomes more important—from managing our critical infrastructures like the electricity grid to managing our private lives—so does the ethics of its technical governance (Cath and Floridi 2017).

The management of the infrastructure of the Internet depends on choice (Lessig 2006) and control (Deibert et al. 2008; Choucri and Clark 2012). It is about how one decides to build the infrastructure through which data travels, and how it does so. The Internet influences who can connect to whom, and how (Denardis 2014). In turn, these choices can have a fundamental impact on the Internet's ability to foster the public interest, especially in terms of social justice, civil liberties, and human rights (Chadwick 2006; Denardis 2014). Control matters too. Internet infrastructure is increasingly becoming 'politics by other means' (Abbate 2000). Understanding the ethics of the practices embedded in the technology underlying the Internet's digital information flows is vital in order to understand and ultimately improve societal and political developments.

To understand this specific axe of digital ethics we need to address professional responsibilities and deontology (Floridi 2012) of those actors involved in coding, maintaining, and updating the Internet's infrastructure, including its applications and platforms. This requires an in-depth understanding of the inner-workings of the Internet. The Internet is not one network but a global network of networks that are bound together through standards and protocols, relying on hardware and software for information flows. Its decentralized, complex, and multi-layered character explains why its maintenance takes many actors and organizations. Considering the complexity of this system, a taxonomy that explains the Internet by dividing it into three distinct layers provides the needed level of abstraction: content, logical, physical (Benkler 2000). This model divides the Internet into three layers: the content layer (information to interact with), the logical infrastructure layer (software), physical infrastructure (wires, cables).

Layer	Description
Content	News, social media posts, videos on streaming platforms, content generated in collaborative tools like Wikipedia or on digital labour platforms
Logical	The technology that makes the Internet interoperable, its digital infrastructure
Physical	The tangible Internet, its physical infrastructure: Computers (servers, personal computers, mobile phones, etc.), telecommunications cables, routers, data centres

Each layer raises specific ethical questions about data, but not all the associated discussions have dedicated institutional homes, and often cut across the various layers and organizations (Mathiason 2008). There has been a concerted political effort over the last decade to highlight how human rights frameworks apply to the Internet. Yet more work remains to be done about how the ethics of data and associated ethics of practices of the various actors mentioned in the taxonomy (Cath and Floridi 2017). There is also limited engagement with the ethical questions surrounding material infrastructures through which data flows. Addressing these questions could alleviate some of the concerns surrounding private ordering of data flows and information control by the Internet's technical and business community, increase trust in the decisions of these private actors (Taddeo and Floridi 2011), and make the overall technical infrastructure of the network more stable, as it combines moral values with an ethical infrastructure to support the instantiation of moral behaviour (Floridi 2012). Such an analytical manoeuvre is particularly important as the infrastructure of the Internet increasingly enacts its social ordering upon the world (Denardis 2014; Hofmann et al. 2016).

2.3   Cyber Conflicts

Cyber conflicts arise from the use of digital technologies for immediate (tactic) or intermediate (strategic) disruptive or destructive purposes. When compared to conventional warfare, cyber conflicts show fundamental differences: their domain ranges from the virtual to the physical; the nature of their actors and targets involves artificial and virtual entities alongside human beings and physical objects; and their level of violence may range from non-violent to potentially highly violent phenomena.

These differences are redefining our understanding of key concepts like harm, violence, combatants, and weapons. They also pose serious ethical and policy problems concerning *risks*, *rights*, and *responsibilities* (the 3R problems) (Taddeo 2012). Start from the risks. Estimates indicate that the cyber security market will be worth US$170 billion by 2020 (Markets and Markets 2015), posing the risk of a progressive weaponisation of cyberspace, which may spark a cyber arms race and competition for digital supremacy, further increasing the possibility of escalation and conflicts (Taddeo 2017a, b; Taddeo and Floridi 2018). At the same time, cyber threats are pervasive. They can target, but can also be launched through, civilian infrastructures. This may (and in some cases already has) initiate policies of higher levels of control, enforced by governments in order to detect and deter possible threats. In these circumstances, individual rights, such as privacy and anonymity, come under devaluing pressure (Arquilla 1999). Ascribing responsibilities is also problematic, because cyber conflicts are increasingly waged through autonomous systems (Cath et al. 2017). Two good examples are the Active Cyber Defence programmes developed in the US and UK, and 'counter autonomy' systems, which are autonomous machine-learning systems able to engage in cyber conflicts by adapting and evolving to deploy and counter ever-changing attack strategies. In both cases, it is unclear who or what is accountable and morally responsible for the actions performed by these systems.

If left unaddressed, the 3R problems will daunt attempts to regulate cyber conflicts, favour escalation, and jeopardize international stability. Digital ethics offers a valuable framework to address the 3R problems, as these span across the conceptual space identified at the beginning of this chapter, namely data, algorithms, and practices.

More specifically, as state-run cyber operations will rely on machine learning and neural network algorithms, focusing on ethics of algorithms will be crucial to mitigate issues concerning the risks of escalation. This is because ethical analyses of algorithms foster the design and deployment of verification, validation, and auditing procedures that can ensure transparency, oversight, on autonomous systems deployed for threat detection and target identification. The ethics of data offers the conceptual basis to solve the friction between cyber conflicts and rights. For it sheds light on the moral stance of digital objects, (artificial) agents, and infrastructures involved in cyber conflicts and, in doing so, it facilitates the application of key principles of Just War Theory—such as proportionality, self- defence, discrimination—to cyber conflicts (Taddeo 2014, 2016). The ethics of practices plays a central role in the regulation of cyber conflicts, as it fosters the understanding of roles and responsibilities of the different stakeholders (private companies, governmental agencies, and citizens) and, thus, shapes the ethical code that should inform their conduct. These problems need to be addressed now, while still nascent, to ensure fair and effective regulations.

2.4   Conclusion

This chapter provided an overview of digital ethics, a new and fast developing field of research that is of vital importance to develop our information societies well, to improve our interactions among ourselves and with our environments, to protect human dignity and to foster human flourishing in the digital age. Much work lies ahead, and progress will require multidisciplinary collaboration among many fields of expertise and a sustained, unflinching focus on the value that ethical thinking can and should make in the world and in technological innovation.

References

Abbate, Janet. 2000. *Inventing the internet.* Cambridge, Mass: The MIT Press.
Arquilla. 1999. Ethics and information warfare. In *Strategic appraisal: The changing role of information in warfare,* ed. Zalmay Khalilzad and John Patrick White, 379–401. Santa Monica, CA: RAND.
Benkler, Yochai. 2000. From consumers to users: Shifting the deeper structures of regulation towards sustainable commons and user access. *Federal Communications Law Journal* 52: 3.
Bynum, Terrell. 2015. Computer and information ethics. In *The Stanford encyclopedia of philosophy.* http://plato.stanford.edu/archives/win2015/entries/ethics-computer/.
Cath, Corinne, and Luciano Floridi. 2017. The design of the internet's architecture by the Internet Engineering Task Force (IETF) and human rights. *Science and Engineering Ethics* 24 (2): 449–468.
Castells, Manuel. 2007. Communication, power and counter-power in the network society. *International Journal of Communication* 1 (1): 238–266.
Cath, Corinne, Sandra Wachter, Brent Mittelstadt, Mariarosaria Taddeo, and Luciano Floridi. 2017. Artificial intelligence and the "Good society": The US, EU, and UK approach. *Science and Engineering Ethics*: 1–24.

Chadwick, Andrew. 2006. *Internet politics: States, citizens, and new communication technologies.* Oxford: Oxford University Press.

Choucri, N., Clark, D. 2012. *Integrating cyberspace and international relations: The Co-Evolution.* Working Paper No. 2012–29, Political Science Department, Massachusetts Institute of Technology. Retrieved from http://ssrn.com/abstract=2178586

Deibert, Ronald. 2008. *Access denied: The practice and policy of global internet filtering.* Cambridge, Mass: MIT Press.

Denardis, Laura. 2014. *The global war for internet governance.* New Haven: Yale University Press.

Floridi, Luciano. 2006. Information ethics, its nature and scope. *SIGCAS Computer Society* 36 (3): 21–36.

———. 2008. The method of levels of abstraction. *Minds and Machines* 18 (3): 303–329.

———. 2011. "The philosophy of information". Oxford/New York: Oxford University Press.

———. 2012. Distributed morality in an information society. *Science and Engineering Ethics* 19 (3): 727–743.

———. 2013. *The ethics of information.* Oxford: Oxford University Press.

———. 2014. Open data, data protection, and group privacy. *Philosophy & Technology* 27 (1): 1–3. https://doi.org/10.1007/s13347-014-0157-8.

———. 2016. Faultless responsibility: on the nature and allocation of moral responsibility for distributed moral actions. *Philosophical Transactions of the Royal Society A* 374 (2083): 20160112.

Hoare, Charles Antony Richard. 1972. Structured programming. In *Structured programming*, ed. O.J. Dahl, E.W. Dijkstra, and C.A.R. Hoare, 83–174. London: Academic. http://dl.acm.org/citation.cfm?id=1243380.1243382.

Hofmann, Jeanette, Christian Katzenbach, and Kirsten Gollatz. 2016. Between coordination and regulation: Finding the governance in internet governance. *New Media & Society* 19 (9): 1406–1423.

Lessig, Lawrence. 2006. *Code: And other Laws of cyberspace, version 2.0.* New York: Basic Books.

Markets and Markets. 2015. 'Cyber security market by solutions & services - 2020'. http://www.marketsandmarkets.com/Market-Reports/cyber-security-market-505.html?golid=CNb6w7mt8 MgCFQoEwwodZVQD-g.

Mathiason, John. 2008. *Internet governance: The new frontier of global institutions.* Routledge.

Mittelstadt, Brent Daniel. 2016. In *The ethics of biomedical big data*, Law, governance and technology series, ed. Luciano Floridi, vol. 29. Cham: Springer.

Moor, James H. 1985. What is computer ethics? *Metaphilosophy* 16 (4): 266–275.

Parker, Donn B. 1968. Rules of ethics in information processing. *Communications of the ACM* 11 (3): 198–201. https://doi.org/10.1145/362929.362987.

Taddeo, Mariarosaria. 2010. Trust in technology: A distinctive and a problematic relation. *Knowledge, Technology & Policy* 23 (3–4): 283–286. https://doi.org/10.1007/s12130-010-9113-9.

———. 2012. Information warfare: A philosophical perspective. *Philosophy and Technology* 25 (1): 105–120.

———. 2014. Just information warfare. *Topoi* 35 (1): 213–214.

———. 2016. On the risks of relying on analogies to understand cyber conflicts. *Minds and Machines* 26 (4): 317–321.

———., ed. 2017a. *The responsibilities of online service providers.* New York/Berlin/Heidelberg: Springer.

———. 2017b. Cyber conflicts and political power in information societies. *Minds and Machines* 27 (2): 265–268.

Taddeo, Mariarosaria, and Luciano Floridi. 2011. The case for E-trust. *Ethics and Information Technology* 13 (1): 1–3.

———. 2015. The debate on the moral responsibilities of online service providers. *Science and Engineering Ethics* (November). https://doi.org/10.1007/s11948-015-9734-1.

———, eds. 2017. *The responsibilities of online service providers*. New York: Springer Berlin Heidelberg.

———. 2018. Regulate artificial intelligence to avert cyber arms race. *Nature* 556 (7701): 296–298. https://doi.org/10.1038/d41586-018-04602-6.

Taylor, Linnet, Luciano Floridi, and Bart van der Sloot, eds. 2017. *Group privacy: New challenges of data technologies*. Hildenberg: Philosophical Studies, Book Series, Springer.

Turilli, Matteo, and Luciano Floridi. 2009. The ethics of information transparency. *Ethics and Information Technology* 11 (2): 105–112.

. . . .
. . . .

Chapter 3
Do We Need a Critical Evaluation of the Role of Mathematics in Data Science?

Patrick Allo

Abstract A sound and effective data ethics requires an independent and mature epistemology of data science. We cannot address the ethical risks associated with data science if we cannot effectively diagnose its epistemological failures, and this is not possible if the outcomes, methods, and foundations of data science are themselves immune to criticism. An epistemology of data science that guards against the unreflective reliance on data science blocks this immunity. Critical evaluations of the epistemic significance of data and of the impact of design-decisions in software engineering already contribute to this enterprise but leave the role of mathematics within data science largely unexamined. In this chapter we take a first step to fill this gap. In a first part, we emphasise how *data*, *code*, and *maths* jointly enable data science, and how they contribute to the epistemic and scientific respectability of data science. This analysis reveals that if we leave out the role of mathematics, we cannot adequately explain how epistemic success in data science is possible. In a second part, we consider the more contentious dual issue: Do explanations of epistemic failures in data science also force us to critically assess the role of *maths* in data science? Here, we argue that mathematics not only contributes mathematical truths to data science, but also substantive epistemic values. If we evaluate these values against a sufficiently broad understanding of what counts as epistemic success and failure, our question should receive a positive answer.

3.1 Introduction

Contemporary data practices are shaped by how mathematical techniques and software turn data into something more valuable, like insight, actionable knowledge, or scientific claims, and develop procedures that make it easier to use data as evidence for specific conclusions, decisions, or actions. This description applies to

P. Allo (✉)
Centre for Logic and Philosophy of Science, Vrije Universiteit Brussel, Brussels, Belgium

Oxford Internet Institute, Digital Ethics Lab, University of Oxford, Oxford, UK
e-mail: patrick.allo@vub.be

© Springer Nature Switzerland AG 2019
C. Öhman, D. Watson (eds.), *The 2018 Yearbook of the Digital Ethics Lab*,
Digital Ethics Lab Yearbook, https://doi.org/10.1007/978-3-030-17152-0_3

algorithms, but also to many ICT-based data practices we find in the sciences. In the latter case, concerns over a lack of human oversight are less salient, but there remains an urgent need to understand the epistemological implications of how a data-centric approach to science reshapes scientific practices.

Through the development of a critical research agenda that aims to understand, contextualise, and evaluate the impact that contemporary data practices have on the creation of knowledge and the justification of decisions, we have become aware of the fact that data is never *just* data, or that the software that is used to process data is never developed in a value-free context. We now accept that there is no such thing as "raw data," (Gitelman 2013), and that design-decisions in software engineering are never exclusively guided by purely technical requirements (Kitchin and Dodge 2011; MacKenzie 2006). As a consequence, there is a growing agreement that the phrase "letting the data speak for themselves" is highly problematic.[1]

It is therefore remarkable that in a context where algorithms have been framed as "weapons of math destruction" (O'Neil 2016), and where the term "math washing" (Benenson 2016) could become synonymous with everything that may go wrong with algorithmic decision-procedures (Mok 2017), surprisingly little has been said about how mathematics – understood as mathematical theories and techniques, but also mathematical thought, and beliefs about the nature of mathematical knowledge – influences the functioning and epistemic status of contemporary data practices. The goal of this paper is to explore this omission, to ask whether this omission is problematic, and (if so) to set out the lines for an expanded research agenda. The challenge this presents to existing critical endeavours is the following: Why, if data are never just data, or code is never just the neutral implementation of a well-defined logic or procedure, do people insist on maintaining that mathematics is *just* mathematics? Is the contribution of mathematics to data science epistemically innocent in a sense that no other pillar of data science could ever claim to be? And if this lack of attention to mathematics is unwarranted, what are the hitherto hidden concerns we could reveal by paying more attention to the role of mathematics in contemporary data practices?

The first two sections are preliminary. I delineate the problem domain, describe the relevant scholarly background, and clarify the relevant concepts. The proposed questioning of mathematical knowledge is retraced to its origins in the sociology of science, and the present inquiry is characterised as the search for (causal) explanations for the scientific respectability of an epistemic practice. Section 3.2 ends by making explicit how this is connected to the development of the field of data ethics. In Sect. 3.3 I further refine the problem statement. In particular, I propose to focus on causal factors that not only increase the credibility of contemporary data practices, but also create obstacles to their critical evaluation.

---

[1] This phrase pops up in very diverse contexts, from the early criticism in human geography by Peter Gould (Gould 1981), to discussions of the choice of uninformative priors in Bayesian statistics (Koop et al. 2007, p. 80), and descriptions of the virtues of large observational studies in astronomy (Katz 2017).

Section 3.4 explicitly develops the analogy between data, code, and maths within data science, and explores how the belief that mathematical knowledge is somehow different makes it easier to dismiss its effects. As a counterpoint to the claim that the role of mathematics is relatively unexamined in the literature, I conclude this section with a discussion of three recent scholarly contributions that directly or indirectly relate mathematics to contemporary data practices.

In Sect. 3.5 I come to the formulation of a concrete hypothesis about how some deeply entrenched views about mathematics and the progress of science contribute to the scientific status of data science. This hypothesis is the view that the reliance on mathematical methods within a given discipline or knowledge practice is an indicator of its scientific maturity, and, conversely, that a resistance to the adoption of such methods is a sign of its immaturity.

3.2   Mathematics and Explaining the Credibility of Data Practices

In their most generic form, the questions I put forward in the introduction are hardly novel. At least since David Bloor, the father of the *Strong Programme in the Sociology of Science*, asked "whether sociology can touch the very heart of mathematical knowledge?" (Bloor 1991, p. 84), the sociology of science has struggled with the problem of how we should reconcile the dominant tendency to explain the functioning of mathematics exclusively in terms of its internal (supposedly impersonal) logic with the demand to explain the growth of mathematical knowledge as we would do for any other human intellectual enterprise. The question of how the development of mathematics can become entangled with specific and hardly neutral socio-technical developments has been extensively studied ever since. Taking the work of Donald MacKenzie as an illustrative example, we find a historical and sociological analysis of the development of statistics in Britain (MacKenzie 1981), the role of mathematics in weapon development (Mackenzie 1990), and more recently the analysis of the Gaussian Copula and its role in the subprime mortgage crisis (MacKenzie and Spears 2014a, b). What emerges from this work, as illustrated in the fragment below, is hardly a picture of mathematics as an impersonal, practically and ideologically neutral enterprise:

> The bricolage involved in the construction of Gaussian copula models thus took place in an uneasy interstitial space in which both practical demands and intellectual – on occasion, perhaps even aesthetic – preferences played important roles. (MacKenzie and Spears 2014b, p. 5)

I propose to focus on two more specific questions to develop my investigation of the relative lack of attention for the role of mathematics. First, how does mathematics contribute to the purported scientific status of contemporary data practices? Second, how, if at all, could this explain how certain data practices can become highly resistant to criticism? We can rephrase this goal in analogy with the motivations of the sociology of science, and thereby make explicit how the two enterprises are related: If our goal is to explain the credibility of a given body of knowledge in given context (Barnes 1982, p. xi) – namely contemporary data practices and their outcomes – what then do mathematics and our beliefs about mathematics contribute to the explanations we seek? The analogy with the sociology of science is, however, imperfect.[2] What I seek are broad epistemological explanations, and these do not always have to be distinctively social explanations of the scientific respectability of data science. More specifically, I seek explanations that refer to beliefs, convictions, arguments, or even publicly expressed views that are external to the theories and disciplines that shape contemporary data practices, as statistics, formal learning theory, and software engineering do.

As I will clarify below, the justification and explanation of the status of data science do not entirely coincide. Consider first the theories and disciplines we readily associate with data science, such as statistics, which supplies the tools we need to quantify the uncertainty of our estimates or predictions. As theoretical foundations, we can refer to them to explain how a given mode of inference or justification can be epistemologically reliable. It is, however, more doubtful that the content of these foundational theories can at the same time satisfactorily account for the respectability and status of these inferential and justificatory practices. As Ian Hacking reminds: "We should not *explain* why some people believe p by saying that p is true, or corresponds to a fact or the facts" (Hacking 1999, p. 81). Here, the truth of p is neither a sufficient nor a necessary condition for believing that p. Similarly, we could justify someone's belief in p on the basis of how likely the truth of p is given her prior knowledge, but we would not in general point to this likelihood as the cause of this belief. Instead, we would require an explanation that may include an enumeration of the evidence that increases the chance of p being true, an account of why a sufficiently high probability is a good reason for belief, and the background knowledge that informs the estimation of the relevant probabilistic claims.

The need for explanations of the epistemological respectability of data science is closely related to the prevalence of the "ethical problems posed by the collection and analysis of large datasets" (Floridi and Taddeo 2016, p. 3), like privacy, transparency, fairness, trust, or responsibility, that are studied in the field of data ethics. In this context, unreliable or untrustworthy data practices are perceived as epistemological *and* ethical problems. Mistakes that have practical consequences can cause harm, but the entanglement of ethics and epistemology in contemporary data practices is not limited to the possibility of mistakes. More fundamental connections between ethical and epistemological dimensions of knowledge practices, like trust and responsibility, deserve dedicated attention. In particular, some ethical risks in contemporary data practices are connected to an unreflective reliance on – perhaps even a deference to – current data practices. Current scholarship identifies several

---

[2] Following Foucault and Hacking, the *genealogical* approach would have been a better point of reference here, but would lead us to far into a purely philosophical territory. The reference to the sociology of science, with its extensive attention to mathematics, is in that respect more convenient.

factors – including the secrecy and opacity of the data processes – that can make data-driven results surprisingly hard to challenge. These are *hard* obstacles to contestation and accountability. Here, I would like to extend this to *soft* obstacles: beliefs about the nature of data practices – like their higher reliability or alleged impartiality – that decrease the credibility of *prima facie* reasonable challenges.[3] The resulting ethical problems are not reducible to the direct effects of mistakes, and are perhaps not even entirely reducible to the practical difficulties we encounter when we try to identify these mistakes, like the problems of opacity and secrecy. Instead, they are related to how the scientific status of data science can render certain challenges less credible or less effective. If we want to diagnose how such uncritical attitudes can emerge, we first need to understand the origins of data science's purported scientific respectability.

3.3   The Problem and the Domain

The term "contemporary data practices" used in the introduction is intendedly vague, and was chosen to bring together a variety of epistemic practices under a single header. It lets us focus on the common features of the values and beliefs that shape, at one end of the spectrum, the adoption of algorithms that guide automated decisions, and, at the other, the development of computationally more demanding data practices within the sciences. As Leonelli (2016) emphasises, the breadth of significant data practices in the sciences should not be reduced to the modes of inference they enable. Data-centric science is, instead, more adequately characterised as "a particular mode of attention within research, within which concerns around data handling take precedence over theoretical questions around the logical implications of [a] given axiom or stipulation" (Leonelli 2016, p. 178). This outlook substantially diverges from descriptions of data science that exclusively emphasise its novelty and potential for unprecedented scientific progress (especially in the context of "Big Data"), and equate data science with data mining and statistical learning.[4]

In the context of the present inquiry, we should not exclusively think of contemporary data practices as modes of inquiry (inferential and information-seeking prac-

---

[3] What I have in mind is analogous to how the current evidence-hierarchies within evidence-based medicine lead to the dismissal of evidence of mechanisms whenever "superior" evidence of correlation (from RCTs) is available (Clarke et al. 2013). Earlier controversies over the value of actuarial methods in medicine highlight a similar phenomenon (Dawes et al. 1989).

[4] In view of this, it remains unclear whether the term "data science" can indiscriminately be applied across diverse contexts of application. One may even doubt whether common uses of the term "data science" successfully designate a unified epistemic practice, rather than a loose collection of methods and techniques (the focus of most textbooks and online tutorials). This is not something that should be cleared up here. For now, it is sufficient to make the open-ended scope of the practices I have in mind explicit, and to highlight the ambiguity of using the terms "data science" and "contemporary data practices" interchangeably.

tices), but also as ways to motivate and defend decisions (justificatory practices), and as ways of organising and valuing knowledge.

Contemporary data practices rest, at least to a first approximation, on mathematical, epistemological, and sociotechnical pillars: (a) Mathematically, they are based on the statistical analysis of regularities, and the ability to discover patterns and/or make predictions through the discovery of such regularities. (b) Epistemologically, they rely on data centrism: the belief that data and access to data are the main requirement for progress in science (Leonelli 2016, p. 176ff).[5] (c) Sociotechnically, they are enabled by increased access to data and computational resources.[6] This initial characterisation of data science places considerable weight on inference, but does not thereby reduce the evaluative practices[7] within data science to the application of criteria that are narrowly associated with the reliability and truth-conduciveness of its inferential procedures.

As a rudimentary account of the epistemic labour that is done by contemporary data practices, these three pillars provide a partial explanation of why certain data practices are:

- Inferentially reliable (the soundness of the mathematical foundation),
- empirically grounded (the compelling force of what Kitchin (2014a, b) describes as new empiricist epistemologies), and
- practically feasible (what is calculable in principle, becomes practically computable).

In the next section, I rely on these three criteria to clarify the respective epistemological roles of data, code, and maths within data science.

The issue I turn to next further develops the thesis that what explains the reliability of a given method does not necessarily explain its credibility.

The question of how mathematics contributes to the purported scientific status of contemporary data practices is, at its core, an epistemological question. The primary issue it raises, however, is not one of demarcation (Is it really science? And how do we know it is really science?). Instead, it asks for a causal explanation of why something becomes knowledge, or why a given practice becomes scientifically respectable. In very general terms, the goal is to explain how modes of inquiry and justification become established. We should, in the words of Hacking, explain how "Styles [of reasoning or of thought] become standards of objectivity" (Hacking

---

[5] Data-centrism is uncontroversial when read as the requirement that claims about natural phenomena require empirical grounding. Critical perspectives on data science have underscored how data-centrism can become associated with empiricist and positivist approaches (Kitchin 2014b, Chapter 8). The proposed analysis does not imply this stronger critical stance.

[6] For instance, this aspect is essential for the recent explosion of neural networks (Chollet 2017, §1.2).

[7] A term borrowed from MacKenzie and Spears (2014b, pp. 5–6), where it refers to "a set of practices, preferences and beliefs concerning how to determine the economic value of financial instruments that is shared by members of multiple organizations" (quoted from the abstract). Here, we can use it more broadly to refer to the practices, preferences and beliefs concerning how to determine the epistemic value of data and data processes.

1992, p. 13).[8] When it comes to contemporary data practices, we should seek explanations for something more specific. I see two reasons for adopting a narrower focus.

A first reason is that it is probably unnecessary to posit the emergence of new styles of reasoning (or modes of inference and justification) in contemporary data science. Instead, we could maintain that as an interdisciplinary practice, data science recombines existing styles,[9] or that as a techno-scientific practice, it merely increases the power and effectiveness of the statistical analysis of regularities.[10] This is confirmed by earlier claims that Big Data is best understood not exclusively in terms of its novelty, but in relation to its historical antecedents, like quantification in bureaucratic settings, the rise of public statistics, and social physics, as well as the positivist epistemologies that define these precursors of contemporary data science (Barnes and Wilson 2014; Rieder and Simon 2017).

An alternative analysis that more eagerly underscores the novelty of contemporary data science remains, however, possible. It is indeed tempting to interpret the shift from explanation to prediction in the discovery and evaluation of statistical regularities (Breiman 2001; Shmueli 2010) as the emergence of a new style of reasoning. I do not wish to evaluate the relative merits of this specific interpretation. I only want to caution that this interpretation singles out an aspect of contemporary data science that is already in focus. This feature is regularly used to criticise the non-scientific nature of (some forms) of data science (Hofman et al. 2017). This is the second reason.

This narrows down our focus on the causal factors that contribute to the status of contemporary data science, and that thereby increase the credibility of its inferential and justificatory practices. A recent line of inquiry in the literature that agrees with this broad description focuses on how technological optimism, blind trust in yet unrealised promises, and full-blown magical thinking pervade the public image of contemporary data science, Big Data, and AI (Elish and boyd 2018). The line I pursue is more conventional, and seeks to understand how mathematics and beliefs about mathematics can foster similar forms of epistemological optimism, and unreflective attitudes. The idea that conceptions of correct reasoning, justification, and indeed objectivity and rationality are not independent from the styles of reasoning in which they are established proves very useful. One of the questions that comes to the fore is how modes of inference and justification can expand their scope and migrate from one knowledge-domain, where their legitimacy is already accepted, to a different one, for which they were not originally designed.

---

[8] Hacking's description continues: "(…) because they get at the truth. But a sentence of that kind is a candidate for truth or falsehood only in the context of the style. Thus styles are in a certain sense 'self-authenticating'. Sentences of the relevant kinds are candidates for truth or for falsehood only when a style of reasoning makes them so" (1992, p. 13).

[9] Indeed, as (Leonelli 2016, §5.2) points out, data practices in the life-sciences are often classificatory rather than based on the discovery and analysis of statistical regularities.

[10] This is one of the styles identified by Crombie, and extensively discussed by Hacking, see (Hacking 1990, 2006).

The underlying concern can be summarised as follows. Data science is a mode of inquiry and justification that has acquired a distinct epistemic status. The reasons for this status are varied, and not all of them are equally well understood. If this status makes it harder to formulate credible challenges against a growing reliance on this mode of inquiry and justification, then it risks turning a generalised reliance into a blind deference. This is ethically and epistemologically problematic, and, notably, is not a risk that can be alleviated with better and more data, improved algorithms, more dependable software, or even responsible and accountable engineering practices (on the latter, see Shneiderman 2016). These are all factors that contribute to a sound reliance on data science, but that do not directly help us to develop a critical or reflective reliance on data science.

The contrast between a sound and a reflective reliance on data science is reminiscent of how Lorraine Daston distinguishes between rationality and the self-critical judgements of reason, but it also applies directly to two distinct uses of mathematics in data science. The first is well understood, and connects mathematical understanding of statistical techniques (and more general statistical literacy) to the ability to detect its misuses. At this level, mathematics fulfils a critical role. The second one is less obvious, and relates beliefs about mathematics to trust in data science as a generally applicable, or distinctively valuable mode of inquiry and justification. At this level, beliefs about mathematics can become an uncritical factor, and can make it harder to formulate credible challenges against the broader adoption of the modes of inquiry and justification of data science.

3.4   A Place for Mathematics?

The current critical research agenda on data science is built on a critical understanding of data and code. In this section, I consider how a comparison of mathematics with data and code supports the view that mathematics should be added to this research agenda. I subsequently illustrate how a number of widely accepted differences between mathematical and non-mathematical knowledge may cast doubt on the proposal that the role of mathematics in data science deserves more critical attention.

The suggestion that data, code, and mathematics jointly enable the modes of inference and justification of data science can be developed with the three epistemic criteria I introduced in the previous section. We can summarise it with the following naive account[11] of what makes data science an epistemically valuable technological possibility. Data science has:

---

[11] Not only does it ignore the intricacies of how data are produced, and the many interventions and practical decisions that precede the analysis of a given data set, it also omits how practical limitations imposed by what is computationally feasible or affordable can increase the gap between the ideal mathematical picture and the algorithm that is actually implemented.

- a claim to epistemic reliability in virtue of the soundness and the rigour of the mathematical basis on which it operates,
- a claim to empirical grounding in virtue of its essential dependence on data, and
- a claim to practical usefulness because we have access to the right resources (data and computational power) to turn the theoretical benefits of the modes of inference we know mathematically into actual benefits.

Although a naive picture, it convincingly shows that *something remains unexplained* if we try to reconstruct the modes of inquiry and justification in data science purely as an interaction between data and code. In essence, we miss the conceptual grip on "probable reasoning" that mathematical results grant to those who understand these results. This much is uncontroversial, but the recognition that "in an algorithmic age (…) mathematics and computer science are coming together" (Danaher et al. 2017, p. 1) does not yet lead to the recognition that *something remains unexamined* if we do not supplement the existing critical examination of data and code with a separate examination of mathematics. The presence of mathematics is widely acknowledged, but is seldom closely examined. This could be a contingent feature of the specific disciplinary context in which the critical research agenda on contemporary data practices is developed.[12] But perhaps there are also certain deeply entrenched beliefs about mathematics that make such a critical evaluation inconceivable.

The contrast between the epistemic justification of data science and the explanation of its credibility (Sect. 3.2), immediately suggests which question we should ask next: What does mathematics do in addition to providing an epistemological justification for the specific modes of inference and justification that are at work in data science? An important cluster of beliefs about the nature of mathematics make it all too easy to reply "nothing" to this question.[13]

This cluster of beliefs relate to the exceptionality of mathematical knowledge and mathematical truth (a topic to which I cannot do justice here).[14] These beliefs support the conclusion that, when compared to data and code, mathematics does not give rise to the risks that motivate the development of a critical research agenda on data science. In brief: in the context of data science, mathematical knowledge is not a cause for concern because it is (in more than one sense) neutral, and its development and application is not easily perceived as the locus of substantial and potentially value-loaded design-decisions, or based on the negotiation of practical trade-offs (Ernest 2016). We find beliefs that support this conclusion in several contexts.

---

[12] This does not have to imply that in this context mathematics is perceived as unproblematic. A more plausible hypothesis is that the impact of mathematics is more likely to be understood in terms of the *positivist epistemologies* to which it can become associated.

[13] The reaction of the French mathematician Jean Dieudonné to the sociology of mathematics (quoted in Van Bendegem (2014)) illustrates this view quite nicely: "To the person who will explain to me why the social setting of the small German courts of the eighteenth century wherein Gauss lived forced him inevitably to occupy himself with the construction of a 17-sided regular polygon, well, to him I will give a chocolate medal" (p. 215).

[14] The textbook challenge to this view remains Lakatos (1976).

First, the exceptionality of mathematics is supported by a central tenet of the philosophy and foundations of mathematics. Mathematical results that deserve to be called theorems rest on deductively valid proofs; proofs that show that the truth of a given statement is inescapable. We can interpret this highly idealised sense in which known mathematical truths are inevitable along two lines. The first one is that, once established, challenges to their truth are futile. The second one is that because proofs have to be (formally) rigorous, their correctness can be verified without access to the personal knowledge of the author of that proof. This ideal picture can be (and has been) challenged along several lines, but their effect on how we perceive mathematics remains important.

Second, this conclusion is supported, but also further problematised, in Porter's reconstruction of the origins of quantification in the modern world (Porter 1995). One of his central conclusions is that the *trust in numbers* that emerged from the nineteenth century onwards should not be seen as a response to an internal scientific need, but rather as a reaction against external pressure, and a growing distrust of personal knowledge, the reliance on expertise and discretion in decision-making processes.[15]

> It is sometimes implied that the drive to make decisions by the numbers is just a matter of engineers doing what comes naturally, the consequence of a marriage of technical knowledge and political power. I have already suggested in relation to American accountants and British actuaries that it was not. Numbers were of course important to both, but in each case the profession insisted on the legitimate and necessary role of expert judgment. Not the experts themselves, but powerful outsiders, worked to simplify regulation by reducing judgment to rules of calculation. (Porter 1995, p. 115)

The primary effect of these views is that when data practices are examined to reveal the contingencies of their constitution, and to challenge the advertised virtues of contemporary data practices (smart and innovative, but also reliable and value-neutral; see Christin 2016, p. 28), the mathematical basis of the computations they rely on remains immune because we tend to situate substantial or value-loaded design decisions outside of mathematics. Critiques of quantification then become critiques that apply to code, or the implemented algorithms that are no longer perceived as mathematical objects, but as specific computational artefacts.

As a counterpoint to the earlier claim that the role of mathematics in data science remains relatively unexamined, I want to draw attention to three recent contributions that bring ideas related to mathematical thought into focus: Christin's examination of mechanical objectivity, McQuillan's proposal to identify a neoplatonic metaphysics within the functioning of data science, and Hildebrandt's defence of the incomputable nature of the human self.

---

[15] Porter's work, and its influence on the critical analysis of data practices suggests that aspects of the role of mathematics in data science can be assessed directly in the context of how data are collected and processed. This is evident when it comes to quantificational practices, but does not need to affect the more general proposal to investigate the effect of mathematical thought, and of beliefs about the nature of mathematical knowledge.

The argument Christin (2016) develops is couched in Daston and Galison's uncovering of objectivity as an epistemological virtue that has a history, and that should not be seen as an absolute and permanent value. Among the historical forms of objectivity they identified, it is only mechanical objectivity that is associated with the absence of human judgement, or the need for value-neutral representations. Christin then develops an analogy between mechanical objectivity as mechanical reproduction through photographic techniques, and the functioning of algorithms that do not require human intervention or that are not clouded by personal biases and prejudices. As she argues, mechanical objectivity is more likely to be adopted and valued as a reaction to some problem than for internal scientific reasons. The claim that "only machines can 'cure' experts from their own subjective weaknesses" (p. 31) suggests that such forms of objectivity are inherently tied to machines. This link to machines can be weakened. As explained above, the absence of judgement is also an ideal of mathematical proof, which has its origin in the foundations of mathematics and the emergence of the mathematical theory of computation. In that perspective, mechanical objectivity should also be seen as an ideal of the inferential and justificatory practices of mathematical thought.

The conjecture in McQuillan (2018) is radically different. He claims that:

Data science can be understood as an echo of the neo-platonism that informed early modern science in the work of Copernicus and Galileo. That is, it resonates with a belief in a hidden mathematical order that is ontologically superior to the one available to our everyday senses. (McQuillan 2018, p. 254)

and emphasises that data science does not simply set out to uncover a hidden mathematical structure, but actively reshapes reality in accordance to this structure. Despite its distinct metaphysical flavour, this suggestion is worth entertaining because it effectively zooms in on a concrete effect of beliefs about the nature of mathematical knowledge. Specifically, McQuillan highlights the view that a mathematical structure that fits the world, or that lets us predict and control events, reveals the true structure of reality. As a metaphysical claim, it is an assertion about the structure of the world, but as an epistemological claim, it is a claim about the superiority of mathematical knowledge and the superiority of gaining knowledge of the world through mathematical analysis and description. The metaphysical claim does not do much explanatory work, and might not even sustain closer scrutiny. Indeed, a recent analysis of the changing role of mathematics in data science (Napoletani et al. 2011, 2014, 2017) develops the thesis that data science is best understood as *agnostic science*: a mode of inquiry that relies on mathematical techniques to answer questions about a given phenomenon on the basis of a large body of data, but where these mathematical techniques are not used to track the structure of the phenomenon. Under this alternative analysis, claims about an ontologically superior knowledge lose their force, but beliefs in the epistemic superiority of a mathematical approach can continue to influence our beliefs about the value of the methods of data science. The evaluation of such beliefs requires a closer examination of how mathematics is used as a tool in data science (Lenhard and Carrier

2017), but also the development of stronger ties with the critical research agenda on contemporary data practices.

In a recent paper, Hildebrandt (2019) claims that "we need an understanding of privacy that is capable of protecting what is uncountable, incalculable or incomputable about individual persons." This view hides (in my re-interpretation) two distinct claims: one about the limits of what is computable, another about the proper understanding of the nature of the entities that serve as inputs for computational processes. The former is the subject of the mathematical theory of computation, whose origins lie in the rigorous characterisation of computation and the discovery of its limits (Boolos et al. 2002, III-IV). The latter emphasises that computational processes take data about individuals – not the individuals themselves – as their input. (In this sense, the phrase "the computed self" should be seen as a category-mistake.) The specific connection Hildebrandt establishes between these two views, is disputable, but the more fundamental question it raises about the conceptual gap between problems and their mathematical expression directs our attention to a different kind of problem. It highlights the fact that when we replace a problem by its mathematical reformulation, we do not only reify its interpretation by mapping it onto a mathematical framework, but we also embed it in a new normative framework that may not respect the norms of the domain in which the problem arose.

3.5   Mature Science

To conclude, I can now formulate a concrete hypothesis about the role of mathematics within data science.

Hypothesis   The presence of mathematically formulated general laws, and the use of mathematical techniques to analyse data, are an essential component of all mature science. The scientific respectability of data science is increased when it is described as a more mature successor of existing modes of inquiry and justification in a given field or context. The dominant role of mathematical techniques in data science contributes to the credibility of strategies that are used to present the reliance on data science methods as a sign of maturity.

This hypothesis draws attention to a powerful, though perhaps also less familiar, argument in favour of the generalised adoption of the modes of inquiry from data science across various domains. The argument is unusual because it does not make claims about unprecedented forms of scientific progress and technological innovation, but only refers to an abstract and hard to dispute scientific ideal.

Most authors who explicitly appeal to a notion of mature science use it as a synonym for *normal science* in the sense of Kuhn (1970), where it primarily refers to agreement on methods and unity of scientific theories.[16] An analysis by Mario

---

[16] In a review of a collection of essays edited by Crombie, Paul Feyerabend characterises Kuhn's position as follows: "The point of view dominating mature science brings about a concentration of

Bunge makes this idea of progress towards a mature science more tangible. He distinguishes between surface or "Baconian" growth, which is characterised by "accumulating, generalizing and systematizing information" from depth or "Newtonian" growth, which refers to the introduction of ideas that transcend the yet available information (Bunge 1968, p. 120). Bunge claims that the maturation of science requires both forms of growth, and notes that while mature science is the ultimate goal of the sciences,[17] most of the non-physical sciences, like the social sciences and humanities, but also the life sciences, remain immature because they have yet to formulate general laws.[18]

Claims about mature science are also used in a more informal sense, and are regularly found in evaluations of the current state of a given discipline.[19] The views expressed in these scholarly contributions are presumably even more revealing of how a given epistemic practice acquires its scientific respectability.

> By the early 1930s, physics was a mature science abounding in universally applicable laws. In comparison, organismic biology was overwhelmingly descriptive and lacked quantitative expressions that could apply to a broad range of animals or plants. (Smil 2000)

> Mathematical reasoning does play an essential role in all areas of computer science which have developed or are developing from an art to a science. Where such reasoning plays little or no role in an area of computer science, that portion of our discipline is still in its infancy and needs the support of mathematical thinking if it is to mature. Large portions of software design, development, and testing are still in this stage. (Ralston and Shaw 1980, p. 69)

These references to the ideal of mature science have two recurrent features. First, they see the emergence of generally applicable laws (which typically have a mathematical formulation) as a hallmark of mature science, and even invite the conclusion that when a knowledge-domain resists mathematical reformulation, this should be seen as a sign of immaturity. Second, they rely on the image of physics as the prototypical example of a mature science.

The development of computational social science (CSS), exemplifies both these features. In this growing field where "digital tools to analyze the rich and interactive lives we lead" (Mann 2016, p. 468), methods from statistical physics and complexity science are used to increase our "understanding of how humans organize and interact in our modern society" (Borge-Holthoefer et al. 2016, p. 1). In this disciplinary context, physics is no longer just a regulative ideal towards which all sciences should strive. Instead, the adoption of the scientific method of physics in the social sciences is now literally true. The association of these developments to the

---

effort. The 'foundations' are now 'taken for granted' and research is directed to the solution of 'more concrete and recondite problems'." (Feyerabend 1964, p. 251).

[17] An important motivation here is the view, notably advocated by Hilary Putnam, that mature sciences converge to the truth. The present argument can proceed without having to rely on such assumptions.

[18] The application of these views to data science requires further refinement, especially in views of the famous slogan on "the end of theory" (Anderson 2008), but these complications do not substantially affect the argumentative force of appealing to the ideal of mature science.

[19] See for instance: Brock (2011); Friedman and Abbas (2003); Shanahan et al. (2005).

maturation of the social sciences, however, long predates the current rise of CSS. As part of a broader discourse on the future of sociology, Randall Collins (1994) asks how sociology could become a high-consensus, rapid-discovery science (as the physical sciences did in the seventeenth century), and explicitly considers the promises of precursors of CSS like microsociology.

This movement, where an alliance to mathematical methods is combined with a more general discourse on the future of a discipline, illustrates the kind of argumentative and rhetorical dynamics that allow views about mathematics to play a role in the motivation and certification of the methods of data science. In particular, by connecting the application of mathematical techniques with the development of a mature science, these arguments initially suggest that the adoption of the data science model of inquiry is consistent with the ultimate goal of science, but they potentially also imply that the refusal to adopt this model of inquiry is incompatible with the progress of science.

Acknowledgements I would like to thank the participants to the "Critical Perspectives on the Role of Mathematics in Data-Science" panel at SPT2017 (Darmstadt, Germany) for discussion on this topic. Additional thanks are due to Karen François and Jean Paul Van Bendegem for feedback, and to David Watson and Carl Öhman for their encouragement and careful editorial work.

This paper would never have been written if I had not, thanks to being a member of the Digital Ethics Lab, become aware of the complex interactions between ethical and epistemological dimensions of contemporary data practices.

References

Anderson, C. 2008. The end of theory: The data deluge makes the scientific method obsolete. *Wired*.

Barnes, B. 1982. *T. S. Kuhn and social science*. London/Basingstoke: MacMillan.

Barnes, T.J., and M.W. Wilson. 2014. Big data, social physics, and spatial analysis: The early years. *Big Data & Society* 1 (1): 205395171453536.

Benenson, F. 2016. *'Mathwashing,' Facebook and the zeitgeist of data worship*. Retrieved from http://technical.ly/brooklyn/2016/06/08/fred-benenson-mathwashing-facebook-data-worship/.

Bloor, D. 1991. *Knowledge and social imagery*. 2nd ed. Chicago: The University of Chicago Press.

Boolos, G., J.P. Burgess, and R.C. Jeffrey. 2002. *Computability and logic*. 4th ed. Cambridge: Cambridge University Press.

Borge-Holthoefer, J., Y. Moreno, and T. Yasseri. 2016. Editorial: At the crossroads: Lessons and challenges in computational social science. *Frontiers in Physics* 4: 37.

Breiman, L. 2001. Statistical modeling: The two cultures. *Statistical Science* 16 (3): 199–231.

Brock, A.C. 2011. Psychology's path towards a mature science: An examination of the myths. *Journal of Theoretical and Philosophical Psychology* 31 (4): 250–257.

Bunge, M. 1968. The maturation of science. In *Problems in the philosophy of science*, ed. I. Lakatos and A. Musgrave, 120–147. Amsterdam: North-Holland.

Chollet, F. 2017. *Deep learning with python*. Shelter Island: Manning Publications.

Christin, A. 2016. From daguerreotypes to algorithms: Machines, expertise, and three forms of objectivity. *SIGCAS Computers and Society* 46 (1): 27–32.

Clarke, B., D. Gillies, P. Illari, F. Russo, and J. Williamson. 2013. The evidence that evidence-based medicine omits. *Preventive Medicine* 57: 745–747.

Collins, R. 1994. Why the social sciences wont become high-consensus, rapid-discovery science. *Sociological Forum* 9 (2): 155–177.

Danaher, J., M.J. Hogan, C. Noone, R. Kennedy, A. Behan, A. De Paor, et al. 2017. Algorithmic governance: Developing a research agenda through the power of collective intelligence. *Big Data & Society* 4 (2): 205395171772655.

Dawes, R.M., D. Faust, and P.E. Meehl. 1989. Clinical versus actuarial judgment. *Science* 243 (4899): 1668–1674.

Elish, M.C., and D. Boyd. 2018. Situating methods in the magic of big data and AI. *Communication Monographs* 85 (1): 57–80. https://doi.org/10.1080/03637751.2017.1375130.

Ernest, P. 2016. Mathematics and values. In *Mathematical cultures. The London meetings 2012–2014*, ed. B. Larvor, 189–214. Cham: Springer International Publishing.

Feyerabend, P. 1964. Review of "scientific change.". *British Journal for the Philosophy of Science* 15 (59): 244–254.

Friedman, C.P., and U.L. Abbas. 2003. Is medical informatics a mature science? A review of measurement practice in outcome studies of clinical systems. *International Journal of Medical Informatics* 69 (2–3): 261–272.

Floridi, L., and M. Taddeo. 2016. What is data-ethics? *Philosophical Transactions of the Royal Society A*. 374 (2083): 1–5.

Gitelman, L., ed. 2013. *Raw data is an oxymoron*. Cambridge, MA: MIT Press.

Gould, P. 1981. Letting the data speak for themselves. *Annals of the Association of American Geographers* 71 (2): 166–176.

Hacking, I. 1990. *The taming of chance*. Cambridge: Cambridge University Press.

———. 1992. Style' for historians and philosophers. *Studies in History and Philosophy of Science Part A* 23 (1): 1–20. https://doi.org/10.1016/0039-3681(92)90024-Z.

———. 1999. *The social construction of what?* Cambridge, MA/London: Harvard University Press.

———. 2006. *The emergence of probability: A philosophical study of early ideas about probability, induction and statistical inference*. Cambridge/New York: Cambridge University Press.

Hildebrandt, Mireille. 2019. Privacy as protection of the incomputable self: From agnostic to agonistic machine learning. *Theoretical Inquiries in Law* 20 (1): 83–121.

Hofman, J.M., A. Sharma, and D.J. Watts. 2017. Prediction and explanation in social systems. *Science* 355 (6324): 486–488.

Katz, N. 2017. *Letting the data speak for themselves: What observations tell us about galaxy formation | SAAO*. Retrieved April 3, 2018, from https://www.saao.ac.za/saao-colloquium/letting-the-data-speak-for-themselves-what-observations-tell-us-about-galaxy-formation/.

Kitchin, R. 2014a. Big data, new epistemologies and paradigm shifts. *Big Data & Society Big Data & Society* 1 (1): 1–12.

———. 2014b. *The data revolution: Big data, open data, data infrastructures and their consequences*. Thousand Oaks: Sage.

Kitchin, R., and M. Dodge. 2011. *Code/space: Software and everyday life*. Cambridge: MIT Press.

Koop, G., D.J. Poirier, and J.L. Tobias. 2007. *Bayesian econometric methods*. Cambridge: Cambridge University Press.

Kuhn, T.S. 1970. *The structure of scientific revolutions. The structure of scientific revolutions*. Chicago: University of Chicago Press.

Lakatos, I. 1976. *Proofs and refutations: The logic of mathematical discovery*. Cambridge: Cambridge University Press.

Lenhard, J., and M. Carrier, eds. 2017. *Mathematics as a tool*. Vol. 327. Cham: Springer International Publishing.

Leonelli, S. 2016. *Data-centric biology*. Chicago: University of Chicago Press.

MacKenzie, D.A. 1981. *Statistics in Britain, 1865–1930: The social construction of scientific knowledge*. Edinburgh: Edinburgh University Press.

———. 1990. *Inventing accuracy: A historical sociology of nuclear missile guidance*. Cambridge: MIT Press.

————. 2006. Computers and the sociology of mathematical proof. In *18 unconventional essays on the nature of mathematics*, ed. R. Hersch, 128–146. New York: Springer.

MacKenzie, D.A., and T. Spears. 2014a. 'A device for being able to book P&L': The organizational embedding of the Gaussian copula. *Social Studies of Science* 44 (3): 418–440.

————. 2014b. 'The formula that killed wall street': The Gaussian copula and modelling practices in investment banking. *Social Studies of Science* 44 (3): 393–417.

Mann, A. 2016. Core concepts: Computational social science. *Proceedings of the National Academy of Sciences of the United States of America* 113 (3): 468–470.

McQuillan, D. 2018. Data science as machinic neoplatonism. *Philosophy & Technology* 31 (2): 253–272.

Mok, K. 2017. *Mathwashing: How algorithms can hide gender and racial biases – The new stack*. Retrieved April 3, 2018, from https://thenewstack.io/hidden-gender-racial-biases-algorithms-can-big-deal/.

Napoletani, D., M. Panza, and D.C. Struppa. 2011. Agnostic science. Towards a philosophy of data analysis. *Foundations of Science* 16 (1): 1–20.

————. 2014. Is big data enough? A reflection on the changing role of mathematics in applications. *Notices of the AMS* 61 (5): 485–490.

————. 2017. *Forcing optimality and Brandt's principle*, 233–251.

O'Neil, C. 2016. *Weapons of math destruction: How big data increases inequality and threatens democracy*. New York: Crown.

Porter, T.M. 1995. *Trust in numbers: The pursuit of objectivity in science and public life*. Princeton: Princeton University Press.

Ralston, A., and M. Shaw. 1980. Curriculum '78 – Is computer science really that unmathematical? *Communications of the ACM* 23 (2): 67–70.

Rieder, G., and J. Simon. 2017. Big data and technology assessment: Research topic or competitor? *Journal of Responsible Innovation* 4: 1–20.

Shanahan, M.J., L.D. Erickson, and D.J. Bauer. 2005. One hundred years of knowing: The changing science of adolescence, 1904 and 2004. *Journal of Research on Adolescence* 15 (4): 383–394.

Shmueli, G. 2010. To explain or to predict? *Statistical Science* 25 (3): 289–310.

Shneiderman, B. 2016. Opinion: The dangers of faulty, biased, or malicious algorithms requires independent oversight. *Proceedings of the National Academy of Sciences of the United States of America* 113 (48): 13538–13540.

Smil, V. 2000. Laying down the law. *Nature* 403: 597.

Van Bendegem, J.P. 2014. The impact of the philosophy of mathematical practice on the philosophy of mathematics. In *Science after the practice turn in the philosophy, history, and social studies of science*, ed. L. Soler, S. Zwart, M. Lynch, and V. Israel-Jost, 215–226. New York: Routledge.

Chapter 4
Using Data from Git and GitHub in Ethnographies of Software Development

Andrew Turner

Abstract Laboratory studies provide the classic examples of ethnography in Science and Technology Studies, however ethnographic methods can also offer insight into scientific and technological development that is geographically dispersed and mediated through computational tools and infrastructures. Sociologists of science studying software development projects are presented with a variety of infrastructures that together constitute the life of a software project. Such infrastructures include mailing lists, wikis, forums, and version control systems. In this chapter I consider how the history of a project's development, as recorded in its version control history (specifically, git and GitHub), can provide a useful resource to STS scholars interested in studying software development. I outline how git and GitHub can be used, what kinds of qualitative and quantitative data can be derived from them, and I evaluate the strengths and limitations of using this kind of data.

4.1 Introduction

Laboratory studies provide the classic examples of ethnography in Science and Technology Studies (STS) (see for example, Knorr-Cetina 1999; Latour and Woolgar 1986; Lynch 1985), however scientific and technological development is frequently characterised by work that is geographically dispersed and mediated through multiple computational tools and infrastructures (Hine 2000). This is particularly true in the case of software development: teams can work effectively without ever being in the same place and contributions to projects (especially projects that are open source) may come from a global community of developers. Moreover, interactions between developers take place via a range of systems specially designed to facilitate distributed collaborative work. One key system in this regard, used by the vast majority of software projects, is the version control system: this is the focus

A. Turner (✉)
National Institute for Health Research, Collaborations for Leadership in Applied Health Research and Care West (NIHR CLAHRC West), University of Bristol, Bristol, UK
e-mail: andrew.turner@bristol.ac.uk

C. Öhman, D. Watson (eds.), *The 2018 Yearbook of the Digital Ethics Lab*, Digital Ethics Lab Yearbook, https://doi.org/10.1007/978-3-030-17152-0_4

of this chapter. Version control systems allow a software project to manage the changes made to code by multiple developers. Crucially, they also contain rich information about the development history of a project, which is a valuable source of evidence for STS scholars studying software development. In this chapter I describe how ethnographic approaches are well-suited to studying the content and dynamics of one particular version control system – git – and an important associated infrastructure – GitHub.

The application of ethnographic methods to highly technologically-mediated phenomena, like software development, has given rise to a variety approaches such as 'digital', 'virtual' and 'trace' ethnography (Murthy 2008; Hine 2000; Geiger and Ribes 2011). Despite the classic focus on laboratory ethnographies, methods in STS have close affinities with these approaches due to the explicit recognition, within STS, of the various identities that technologies have and the work that is done to negotiate, stabilise, and destabilise those identities in different contexts. Importantly, the approaches above seek to disrupt the traditional association of ethnographic field sites with particular locations (like laboratories), by aiming to "despatialize notions of community, and focus on cultural process rather than physical place" (Hine 2000, p. 61) (See also: Beaulieu 2010; Jarzabkowski et al. 2015; Mackenzie 2003). For example, Beaulieu has argued that "the ethnographic approach must adapt in order to study these fields in which research practices are not concentrated in lab-like spaces" (2010, p. 456) and that "space, texts and infrastructures become so many resources in establishing co-presence that can be embraced as constitutive of the field" (2010, p. 458). That is to say, one should not aim for researchers to be co-located with their objects of study in a physical sense, but rather they should aim to achieve 'co-presence' in the various systems and infrastructures that constitute the sites of science and technology development. I suggest that a version control system, specifically git and GitHub, is precisely such a resource for studying software development and is an infrastructure where it is relatively easy for STS scholars to achieve co-presence with a project's developers.

As a way to examine the various different situations that may be implicated when studying a particular phenomenon 'multi-site ethnography' is a method that follows people and objects across contexts (Marcus 1995) and aims to provide a characterisation that is more sensitive to the connections and transformations that take place between them. In this sense then, git and GitHub may seem to be simply one more field site in which STS scholars should embed themselves when studying a software development project. However, as Hine notes, the multi-sited approach assumes there are well-defined field sites that researchers move between (Hine 2000, 2007). In contrast, she argues that interactions taking place through various computational systems and infrastructures are fundamentally characterised by "flow and connectivity" rather than "location and boundary". Moreover she suggests that the very notion of a field site is challenged when "the ethnographic object itself can be reformulated with each decision to follow either yet another connection or retrace steps to a previous point" (Hine 2000, p. 64). Hence it is not simply that ethnographies of technologically-mediated phenomena, like software development, must take account of the multiple sites and virtual spaces where science and technology development can take place, but rather that the mediation through those systems and infrastructure throws into sharper relief (by removing the seeming naturalness of spatial definitions) the fact that the boundaries of a field site are always analytical choices and not given *a priori* (Hine 2007). As will be explained below, the decision of 'where to stop' when examining data contained in git and GitHub, and the ambiguity around what constitutes 'the project' within these systems, raises exactly this challenge.

Another innovation in ethnographic methods that de-emphasises space and location and instead explores flow and connectivity is 'trace ethnography' (Geiger and Ribes 2011). Trace ethnography "exploits the proliferation of documents and documentary traces in [...] highly technologically-mediated systems" (Geiger and Ribes 2011, p. 1). Indeed, traces are not just prolific in such systems but also the primary means by which actors interact. Studying these traces promises to illuminate how actors understand such systems and how the texts they produce do not simply document activity but also coordinate it (Geiger and Ribes 2011; Hine 2000). Geiger and Ribes explain how "sets of such documentary traces can then be assembled into rich narratives of interaction allowing researchers to carefully follow coordination practices, information flows, situated routines and other social and organisational phenomena" (Geiger and Ribes 2011, p. 1). This is what studying version control systems promises to deliver.

The direct relevance of Geiger and Ribes's trace ethnography to studying software development is clear in their study of editing practices on Wikipedia. Wikipedia is a platform which keeps a history of changes as content is edited, added or removed. This version history of content on Wikipedia shares much in common with the version history of software code created by git and presented on GitHub. Indeed, Geiger and Ribes offer a tantalising hint towards the applicability of their trace ethnography to version control practices in software development when they note that: "software code repositories often keep detailed records of who changed what and when [and, as described below, why], and are often used to keep developers accountable and let maintainers know how the project has changed at a glance" (Geiger and Ribes 2011, p. 9). The aim of this chapter is therefore to explain how git and GitHub can provide such evidence.

Outside of STS, ethnographic methods have also been applied in the Human-Computer Interaction (HCI) field, focusing for example on the design of user interfaces and user experience (for an exploration of the crossover of STS and HCI, see especially Woolgar 1990). Relatively little attention has been given to ethnographies of the practice of software development, but this is not to say that computer science has ignored the different ways that traces and texts produced within infrastructures like git and GitHub can be used as data sources. Notably, Bird et al. (2009) and Kalliamvakou et al. (2015) have examined the 'promises and perils' of mining git and GitHub, respectively. In much of this literature, however, the focus is on investigating the behaviour of developers or software projects in aggregate, that is to say, looking at git and GitHub usage across projects, or at networks of developers, rather than examining software development at the micro-level and focusing on one specific project, as is more typical of in-depth ethnographic work.

In what follows, this chapter first describes git and GitHub and their combined role in keeping software code under version control, in particular the key notions of a 'commit' and the 'commit history' are introduced. The chapter then explains the features of commits and the commit history that make them useful for studying software development projects. Throughout, the examples of code on GitHub are taken from code from the git project itself, since it is both mature and popular. The chapter concludes by reflecting on the limitations of using commits and the commit history as data for ethnographic research.

4.2   Git and GitHub

4.2.1   What Are Git and GitHub?

Git is a version control system typically used for keeping track of changes that have been made to software code (Git 2018a). A folder of files managed by git is called a *git repository*. Tracking and controlling changes made to code within a git repository makes development more efficient and facilitates collaboration by making management of different versions and potential conflicts easier.

Git is one of many version control systems, each with advantages and disadvantages in particular use-cases. However, git is one of the most popular systems in use in part because of the popularity of a related service: GitHub (https://github.com). GitHub is an online service that offers hosting of git repositories and is one of, if not the, largest website for hosting open source projects (GitHub 2018a). It provides a public and central site where software projects can store their code, manage contributions from other developers, and administer other aspects of the project such as requests for new features and reports of problems. Git, on its own, is a command-line tool and GitHub complements git by presenting the content of git repositories (that is, the code and version history) in a browsable form on the web. Beyond simply presenting the content of git repositories, GitHub also adds its own suite of features on top of git's functionality such as wikis and issue trackers, allowing users to report problems, and allowing anyone interested in the project to suggest new features or propose changes to the code. GitHub gives a software project a central, public, home on the web, and additionally provides a social platform for its users (see Bryan 2017).

This centralising aspect of GitHub and the extra features it adds for users to interact with git repositories and other GitHub users has led to novel uses of GitHub that go beyond keeping code under version control. For example, Kalliamvakou et al. (2015) found repositories on GitHub serving purposes besides software development, including: demonstration code and tutorials; websites and blogs; teaching materials for projects; and personal files such as presentations and software configuration files. However, these other uses are not the focus of this chapter.

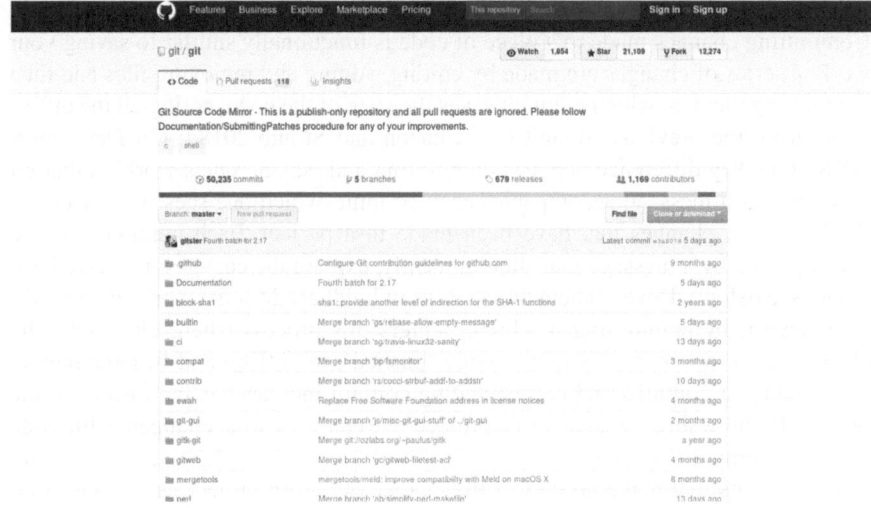

Fig. 4.1 GitHub web interface

Insofar as git and GitHub are used for software development, a key reason git is useful to developers is also a reason why it can be useful to those studying software development. As GitHub describe in their 'git handbook':

> Git lets developers see the entire timeline of their changes, decisions, and progression of any project in one place. From the moment they access the history of a project, the developer has all the context they need to understand it and start contributing. (GitHub 2018b)

Git and GitHub contain rich information about the history and decision-making of a software project and about the social interactions between developers. The combination of git and GitHub introduces a new tool (git) and a new infrastructure (git repositories hosted on GitHub) into software projects that 'materialize' (Latour and Woolgar 1986) the code and development history. By imposing more rigid workflows, git creates the mechanisms for users to visibly document their work and package it in ways that make it easy to distribute to collaborators, and, crucially for this chapter, make it accessible for ethnographic research. To illustrate the appearance of a repository on GitHub, Fig. 4.1 shows the GitHub web interface for the git project.

4.2.2   Commits: The Basic Unit of Version Control

At the simplest level, version control involves being able to save your work and to record which changes took place between saves. This might be achieved manually by incrementing version numbers in a file name, or by using features such as 'Track Changes' to highlight the differences between the original and the edited text in a document. Tracking changes in git is achieved by 'committing' those changes. Committing changes made to a piece of code is functionally similar to saving your work: a series of changes are made by editing, adding and removing files and then committing them, at which point git saves the state of the code, noting all the differences from the previous commit (see Chacon and Straub 2018d; Git Developers 2018). One key difference between committing and 'saving one's work' is that git requires that a message accompanies each commit, which one sees when viewing the history of changes that have been made in a project. Each commit must be accompanied by a message that should, ideally, explain the changes that have been made (see below). Development proceeds as individuals do some work and periodically save it by committing it. GitHub enters this process when, ultimately, the changes are sent, or in git's terminology 'pushed', to GitHub and become public. The totality of commits and corresponding commit messages is the history of the project. By browsing the history of a project, one can view what changed in the code at each commit and read a description of this change. For example, Fig. 4.2 shows some commits that make up part of the history of the git project, as displayed on GitHub.

With this basic understanding of git and how the history is displayed by GitHub, we can explore what kind of ethnographic evidence this provides for STS scholars studying software projects.

4.3   Commits as Data

4.3.1   The Content of Commit Messages

One of the most important aspects of a commit is that it must be accompanied by a message that, ideally, describes the changes that have been introduced since the previous commit. What and when to commit is a choice that developers repeatedly make as they work. It is up to developers to decide what to write for their commit messages and therefore how to describe the changes made in a particular commit. In this way committing can be more purposeful and structured than regularly 'saving'.

Within software development communities, and in fact within the documentation written by git's authors, the practice of committing changes is described as being governed by norms for the accurate description of the changes that have been made (Beams 2014; Chacon and Straub 2018c; Git Developers 2018). For example, git's documentation provides the following guidance on how a 'good' commit message should be structured:

> Though not required, it's a good idea to begin the commit message with a single short (less than 50 character) line summarizing the change, followed by a blank line and then a more thorough description. (Git Developers 2018)

Fig. 4.2 Partial commit history of the git project

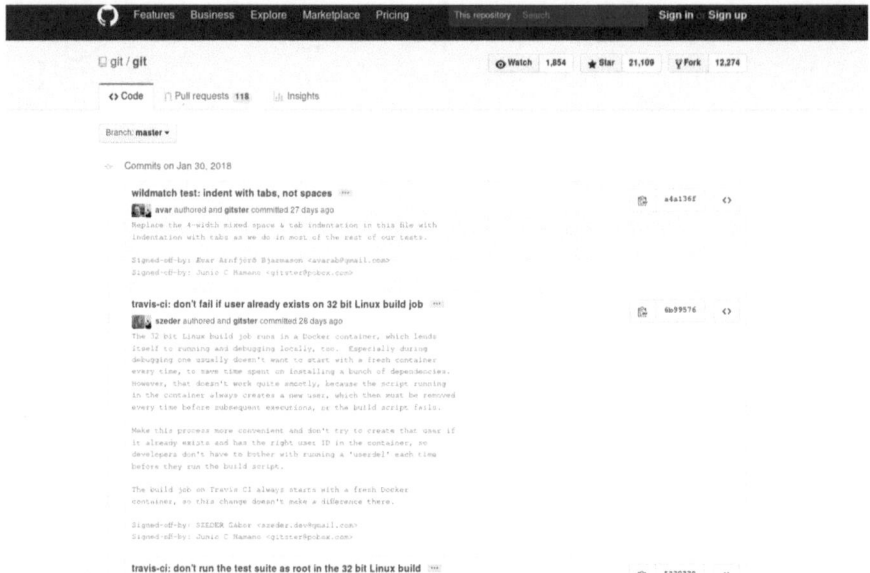

Fig. 4.3 Commit messages within the git project

Moreover, the guidelines for developers contributing to git itself state that the content of commit messages should:

> Give an explanation for the change(s) that is detailed enough so that people can judge if it is good thing to do, without reading the actual patch text [that is, the actual changes to the code] to determine how well the code does what the explanation promises to do [...] Descriptions that summarize the point in the subject well, and describe the motivation for the change, the approach taken by the change, and if relevant how this differs substantially from the prior version, are all good things to have. (Git 2018b)

Commit messages are written by developers to help other developers understand the work they have done and to provide context for the changes, without necessarily requiring others to read code. In this sense, commit messages should translate the code changes into 'plain' language (that is, at least, language that other developers with content knowledge of the project can interpret). Commit messages therefore potentially provide rich insight into *why* changes were made, in the developer's own words. To illustrate consider examples from the git project, where a detailed commit message may appear as shown in Fig. 4.3.

Figure 4.3 shows detailed descriptions accompanying code changes. Taking the content of such commit messages as data, STS scholars can gain insight into what is being worked on or prioritised, why certain design decisions were taken, as well as how developers solve problems related to, say, unforeseen issues with the code that led to bugs, or how developers creatively navigate and manage old code in order to add new features without introducing new bugs or breakages. Commit messages provide a description of all the incremental changes that have been made, and they are descriptions written close to the time that the code changes were made, thereby providing data that is closer to a journal or diary written for others, rather than an interview or survey.

Commit messages are not only ethnographically valuable because of their content, however. Other features of commit messages provide insight into the norms and practices that are collectively endorsed by software projects. Moreover, these commit practices show how a project's contributor guidelines, if such guidelines exist, are in fact translated into practice. That is to say, by viewing the commit messages of a project we can learn about the project's norms for software development: what standards for commit messages are enforced and what additional features are taken advantage of.

To take one example, an additional feature of git is the ability to add tags at the end of commit messages such as 'Signed-off-by' 'Reported-by', 'Tested-by', or 'Reviewed-by' followed by a name and email address. (The 'signed-off-by' tags are visible in Fig. 4.3, for example.) This provides an additional way to track who authored or reviewed a particular commit, but also provides a formal way of recording reviewing, auditing or governance processes that are built into the development and software release process. As noted above, these processes would also otherwise only be visible in contributor guidelines and it would be unclear how they translated into practice. Here again then, the commit history materialises social interactions between developers and norms of the project providing a further way for them to become accessible to STS scholars.

In addition to the messages that accompany each commit, each commit by definition shows the changes made to the code since the previous commit. While the number of files and number of lines of code that have changed is an imperfect indicator of the complexity of a particular change, it does however illustrate the magnitude of the changes that a commit introduces. For example, it can illustrate what needed to be done to implement a new feature or fix a problem, and shows which commits involved the greatest manipulation of the existing code. In quantitative terms, therefore, the commit history of a project shows how much code has changed when, for example, a feature has been implemented or a problem has been fixed.

In fact, the commit history of a project is a source of much quantitative information that can be useful to STS scholars. Metadata that is part of every commit includes the date and time of the commit and the author of the commit. (The 'author' is, in this sense, the person who made the commit. As noted above, extra content in the commit message about who has signed-off, reviewed, etc., the commit provides greater detail about the interactions between developers that led to the final commit.) The frequency of the commits provides insight into the level of activity within a git repository. Indeed GitHub visualises this information as part of the standard way that repositories are displayed. One can, for example, view the number of commits made over time (see Fig. 4.4). This is also broken down by developer, allowing one to see who are the most active contributors and the distribution of their contributions over the commit history.

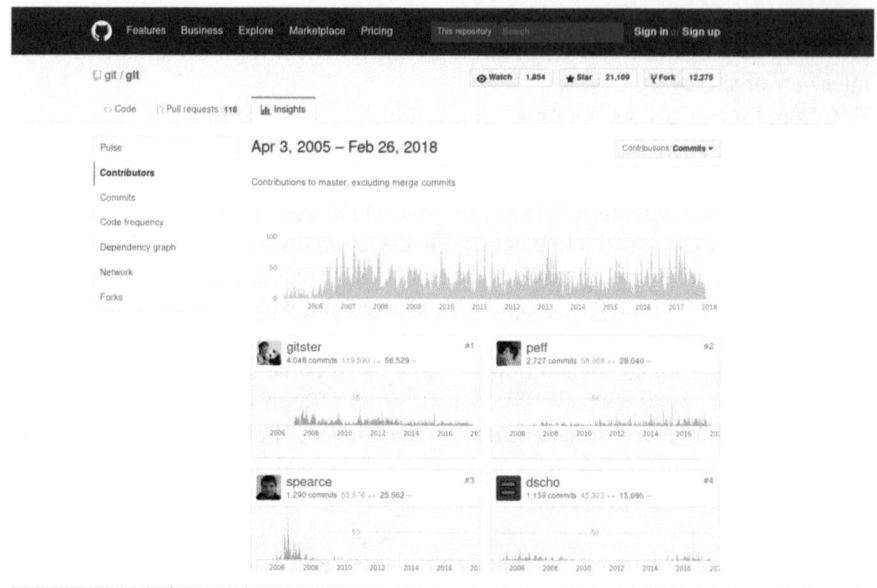

Fig. 4.4 Git contributor information

4.3.2   Where to Find Commits: Branches and Forks

A further important feature of git is 'branching' (with its complement, 'merging') (Chacon and Straub 2018a). There are a range of common but different ways in which software projects can use git and GitHub to implement branch-based development workflows (Driessen 2010; Atlassian 2016; Chacon and Straub 2018b; GitHub 2013). Fundamentally, however, branching allows developers to create independent diverging versions of the code within a repository. For example, maintaining a stable and extensively tested 'master' version of the code alongside various 'development' versions that contain new features or work in progress. Branches provide a way to maintain divergent versions of code without creating unmanageable duplication. Moreover, they provide a space for experimental or task-specific work to take place, without disrupting the 'master' version. Git offers mechanisms for merging code between branches that makes the integration of these parallel work-streams relatively simple (One can see evidence of extensive merging in Fig. 4.2). Branches materialise and make visible the different work-streams that are going on in a project, providing insight into how the code is put together to create each new release of the software. Figure 4.5 illustrates the different public branches that exist in the git project.

The existence of multiple branches in a project crucially means that there is not a *single* commit history; rather, each branch represents a parallel history from the point it diverged. Ultimately of course, one would expect a branch to be merged back into the master branch when, for example, the implementation of a new feature

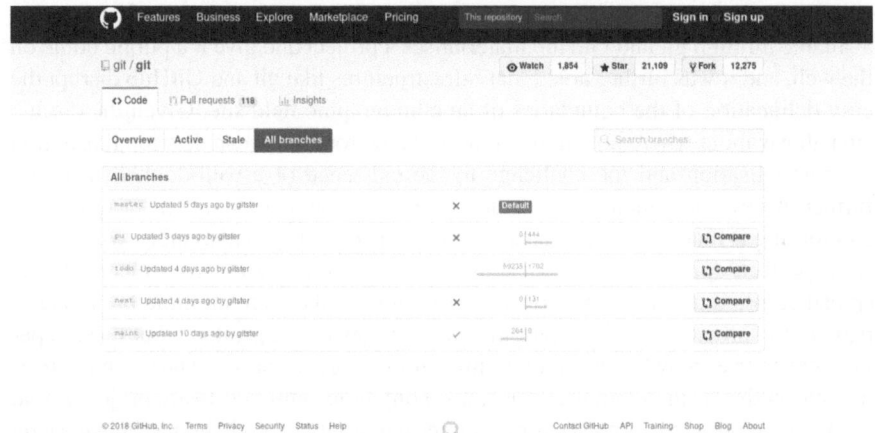

Fig. 4.5 Git branch information

had been completed, or a problem fixed. For the STS scholar, however, the key consequence of a project using multiple branches is that commits may exist on one branch but not another, and therefore analysis of commits on the master branch is insufficient to capture the totality of commits.

A subtler, related point to note is that the commits that are publicly visible on GitHub may not represent the totality of a developer's work. For example, developers may maintain their own *private* branches, which they never make public via GitHub. Git is a distributed version control system and STS scholars should not assume that what is visible on GitHub represents the entirety of what is kept under version control. Indeed, there may be significant value in working with individual developers to gain access to their private branches that provide greater insight into how development proceeds and into how they prepare their code to a point where they are happy to share it publicly.

This same issue about where to find potential commits to a project's code is also instantiated in another way. GitHub makes a feature known as 'forking' available to users. Forking means making a personal copy of a repository so that it can be modified: any user can fork a repository (i.e. create their own personal copy of the whole repository) whereas only users authorised by the project in question can create branches (i.e. create a new work-stream *within* the repository). The complement to forking is a 'pull-request' which is a way to request that changes made in a forked copy of the code should be integrated in the original repository. In many ways a fork and pull-request is functionally similar to a branch and merge. An important methodological point from the perspective of someone studying software development on git and GitHub is that new commits can be spread across branches within a repository (as noted above), and they can be spread across other repositories that were forked from the main repository. Commits made in those forked repositories may be waiting to be – in GitHub's language – 'pulled' into the main repository.

In the introduction it was noted that making code and the history of its changes available through git and GitHub materialises a project and give it a public home on the web, and it was further noted that infrastructures like git and GitHub disrupt the easy delineation of the boundaries of an ethnographic field site. Given the discussion above about where potential commits can be found, we can develop this notion of materialisation and the challenge to the existence of a well-defined field site further. A key consequence (also noted by Kalliamvakou et al. 2015) of branching and forking is that a single git repository is not necessarily synonymous with a software project. An obvious reason is that the code for a software project may be spread across multiple repositories. For example, if there are multiple but modular parts to the project, then each part may have its own git repository (but these separate repositories may be owned by a single user or organisation). The more interesting reason that a single repository is not synonymous with a software project is that work on the project is not necessarily represented only by code in the repository *on GitHub*. Rather, as we have seen above, code can exist on private branches that developers never push to GitHub; equally it can exist publicly on GitHub, but contained in forks of the main repository (and of course there may be private branches that belong to those forks, too). As a result, the question of what a software project is, in terms of its footprint within git and GitHub and extending beyond that infrastructure, is a key question for STS scholars studying software development and an important consideration when evaluating the scope of the site that GitHub offers.

4.3.3 Interpreting the Commit History

A major limitation of treating commit messages as data is that software projects may simply not produce helpful commit messages. The examples given above, from the git project itself, are good examples because git is a mature, well organised project where norms for writing detailed commit messages and ensuring a tidy commit history are internalised and enforced. In contrast, other projects may have norms and practices for using git and GitHub that do not generate the same rich development history. To give three examples: first, commit messages may be terse or poor descriptions of the changes that have been made. Second, the changes introduced within each commit may be an accumulation of unrelated changes. Third, a software project may not use git and GitHub as a platform for collaboration on code, but instead use it purely as a platform for publicly releasing their code and therefore not publish a development history at all. That is to say, there may only be a single commit for each new release of the software. Again, the possibility of these scenarios reinforces the point that the commit history of a project is actively crafted by developers according to norms for transparent and collaborative development, indeed, this is a large part of what makes it valuable as data.

It is important to understand the degree to which a commit history can be crafted by developers, because this affects interpretation. Specifically, it is important to appreciate that history can be changed and that commits visible on GitHub do not necessarily reflect the way the work was performed. For instance, git allows developers to 'tidy up' their work before pushing it to GitHub. This can include a range of manipulations such as re-ordering commits (for example, to create a more logical and less chronological history), adding, removing or merging commits (for example, to carve up the history into different, perhaps more functionally discrete, chunks), while at the same time re-writing commit messages so they describe a more coherent or purposeful history of the changes.

The fact that commit histories appear as objective logs of the work that has been done should not obscure the fact that they are created, and re-created, by developers making choices about the 'tidiest' or 'most logical' way to present their work to other developers. This applies to commits *before* they are pushed to GitHub. While it is possible to change history *after* pushing to GitHub (that is, after the commits have become public) this can result in significant problems for other developers and git has mechanisms to ensure one cannot do this by accident. Consequently, the commit history visible on GitHub should not be seen as history-as-it-happened, but rather the history-that-a-developer-wants-to-tell. A related methodological challenge highlighted by Bird et al. (2009) is that the order that commits appear in the history can be misleading and it can be difficult to identify which branches they originate from. This cautions against attempts to reconstruct, from the commit history of a project, which branches were merged at which points. However, this is an issue about the difficulty of establishing the chronology of the history, rather than interpretation.

Nevertheless, it is also crucial to note that the commit history of a project is not only useful because of its content. As the above points suggest and as Geiger and Ribes note, there is a worry that such histories can be "notoriously 'thin' and on the surface, often [they] do not appear to hold much evidence about the actions they describe" however, as they go on to argue, their value is not solely tied to their *content* and "the utility of such traces does not stem from some inherent documentary quality, but rather because they are produced and circulated within a highly standardized sociotechnical infrastructure of documentary practices" (2011, p. 5).

4.4 Conclusion

Git and GitHub present a wealth of information about how a software development project is conducted. They make the design choices and priorities of developers and the dynamics of collaboration accessible in new ways. This chapter has described some of the ways in which such information is materialised by git and GitHub in the commit history. Of course, the data provided by these infrastructures must be interpreted cautiously. The commit messages and commit history are textual data just like other qualitative data and are written to achieve and perform software development just as much as they seem to passively log and document development activities. Also, software projects use git and GitHub in different ways, according to . different norms, which may restrict the richness and volume of information available. For example, repositories may have histories made up of short and mini-mally descriptive commit messages, and projects may only engage in sharing the very minimum amount of finished code rather than rich histories of development. In either case, this does not necessarily limit the usefulness of studying such projects through git and GitHub, although it does reduce the 'thickness' of the content of the version control history. Rather the ways in which git and GitHub are used reveals the norms of a software project and the extent to which a project engages with the many ways it can allow itself to be mediated through those infrastructures. Crucially, git and GitHub provide qualitative data about software projects that is either not readily accessible elsewhere or not materialised at all. It complements and can be triangulated with the kinds of evidence provided by other infrastructures that also constitute the life of a software project.

References

Atlassian. 2016. *Forking workflow*. Atlassian git tutorial. https://www.atlassian.com/git/tutorials/comparing-workflows. Accessed 9 Mar 2018.

Beams, C. 2014. *How to write a git commit message*. https://chris.beams.io/posts/git-commit/. Accessed 13 Feb 2018.

Beaulieu, A. 2010. Research note: From co-location to co-presence: Shifts in the use of eth-nography for the study of knowledge. *Social Studies of Science* 40 (3): 453–470. https://doi.org/10.1177/0306312709359219.

Bird, C., P.C. Rigby, E.T. Barr, D.J. Hamilton, D.M. German, and P. Devanbu. 2009. The promises and perils of mining git. In *6th IEEE international working conference on mining software repositories, 2009*. MSR 2009, 1–10. https://doi.org/10.1109/MSR.2009.5069475.

Bryan, J. 2017. Excuse me, do you have a moment to talk about version control? *The American Statistician* 72: 20–27. https://doi.org/10.1080/00031305.2017.1399928.

Chacon, S., and B. Straub. 2018a. *Branches in a nutshell*. Git-Scm. https://git-scm.com/book/en/v2/Git-Branching-Branches-in-a-Nutshell. Accessed 13 Feb 2018.

———. 2018b. *Branching workflows*. Git-Scm. https://www.git-scm.com/book/en/v2/Git-Branching-Branching-Workflows. Accessed 9 Mar 2016.

———. 2018c. *Contributing to a project*. Git-Scm. https://www.git-scm.com/book/en/v2/Distributed-Git-Contributing-to-a-Project#_commit_guidelines. Accessed 30 Jan 2018.

———. 2018d. *Recording changes to the repository*. Git-Scm. https://git-scm.com/book/en/v2/Git-Basics-Recording-Changes-to-the-Repository. Accessed 13 Feb 2018.

Driessen, V. 2010. *A successful git branching model*. Nvie.com. http://nvie.com/posts/a-success-ful-git-branching-model/. Accessed 5 Jan 2010.

Geiger, R.S., and D. Ribes. 2011. Trace ethnography: Following coordination through documen-tary practices. In *System sciences (HICSS), 44th Hawaii international conference on*, 1–10. IEEE.

Git. 2018a. *Git*. Git-Scm. https://git-scm.com/. Accessed 16 Feb 2018.

———. 2018b. *Git: Documentation/SubmittingPatches*. https://github.com/git/git/blob/master/Documentation/SubmittingPatches. Accessed 13 Feb 2018.

Git Developers. 2018. *Git-commit documentation*. Git-Scm. https://git-scm.com/docs/git-commit.

GitHub. 2013. *Understanding the GitHub flow*. GitHub guides. https://guides.github.com/intro-duction/flow/index.html. Accessed 13 Feb 2018.

———. 2018a. Build software better, together. GitHub. https://github.com/about. Accessed 13 Feb 2018.

———. 2018b. *Git handbook*. GitHub guides. https://guides.github.com/introduction/git-handbook/. Accessed 13 Feb 2018.

Hine, C. 2000. *Virtual ethnography*. London: Sage.

———. 2007. Multi-sited ethnography as a middle range methodology for contemporary STS. *Science, Technology & Human Values* 32 (6): 652–671. https://doi.org/10.1177/0162243907303598.

Jarzabkowski, P., R. Bednarek, and L. Cabantous. 2015. Conducting global team-based ethnography: Methodological challenges and practical methods. *Human Relations* 68 (1): 3–33. https://doi.org/10.1177/0018726714535449.

Kalliamvakou, E., G. Gousios, K. Blincoe, L. Singer, D.M. German, and D. Damian. 2015. An in-depth study of the promises and perils of mining GitHub. *Empirical Software Engineering* 21: 2035–2071. https://doi.org/10.1007/s10664-015-9393-5.

Knorr-Cetina, K. 1999. *Epistemic cultures: How the sciences make knowledge*. London: Harvard University Press.

Latour, B., and S. Woolgar. 1986. *Laboratory life: The construction of scientific facts*. Chichester: Princeton University Press.

Lynch, M. 1985. *Art and artifact in laboratory science: A study of shop work and shop talk in a research laboratory*. London: Routledge & Kegan Paul.

Mackenzie, A. 2003. These things called systems: Collective imaginings and infrastructural software. *Social Studies of Science* 33 (3): 365–387. https://doi.org/10.1177/03063127030333003.

Marcus, G.E. 1995. Ethnography in/of the world system: The emergence of multi-sited ethnography. *Annual Review of Anthropology* 24 (1): 95–117. https://doi.org/10.1146/annurev.an.24.100195.000523.

Murthy, D. 2008. Digital ethnography: An examination of the use of new technologies for social research. *Sociology* 42 (5): 837–855. https://doi.org/10.1177/0038038508094565.

Woolgar, S. 1990. Configuring the user: The case of usability trials. *The Sociological Review* 38 (S1): 58–99. https://doi.org/10.1111/j.1467-954X.1990.tb03349.x.

Chapter 5
The Price of Discovery: A Model of Scientific Research Markets

David Watson

Abstract Large scale scientific projects often require enormous investment on the part of their patrons to develop and maintain new technologies like particle accelerators and grid computing systems. Data suggest that funding for such research is increasingly concentrated into a small number of major grants awarded to just a handful of successful labs. But is this the most efficient market design for promoting discovery? In this chapter, I propose a modified solution to the monopsony profit maximisation problem that specifies the conditions under which patrons should invest in one or several labs for a given research project. I apply the model to two prominent case studies and identify the key indicators that policymakers should consider when deciding how best to fund new scientific research.

5.1 Introduction

It is a well-documented fact that information and technology markets tend to be dominated by monopolies (Klemperer 2008; Shapiro and Varian 1998; Varian et al. 2004). This was true long before the advent of the internet. AT&T, Paramount Pictures, CBS, and a handful of other firms all took their turn at the throne of their respective information industries (Wu 2010). The tradition continues today on the World Wide Web, where Google, Amazon, and Facebook rule their digital dominions with every bit as much authority as their old media forebears exerted last century.

The same forces that drove those markets toward monopolisation have had similar effects on a different, less widely discussed sector of the economy: scientific research. As information technology plays an increasingly important role in cutting edge work across the natural sciences, funding has gradually gravitated toward a small number of labs with an outsized influence on the development of their fields. This shift has major epistemological and ethical implications, as opportunities for

D. Watson (✉)
Oxford Internet Institute, Digital Ethics Lab, University of Oxford, Oxford, UK
e-mail: david.watson@oii.ox.ac.uk

© Springer Nature Switzerland AG 2019　　　　　　　　　　　　　　　　　　51
C. Öhman, D. Watson (eds.), *The 2018 Yearbook of the Digital Ethics Lab*,
Digital Ethics Lab Yearbook, https://doi.org/10.1007/978-3-030-17152-0_5

innovation are concentrated in a small and shrinking community of researchers. While certain factors clearly distinguish scientific research from more conventional goods and services, I argue that in both types of markets, similar patterns of production and consumption have led to essentially the same results. In both cases, a brief period of fierce competition is followed by the sustained domination of a few powerful agents. Whether such an outcome is efficient or not depends on a variety of factors.

In this chapter, I critically examine these factors, and propose a model of modern scientific research markets. Section 5.2 lays the groundwork for this model, elaborating upon the ways in which the production and consumption patterns of scientific research resemble and differ from those of more familiar information and technology markets. In Sect. 5.3 I formalise the model's components, assumptions, and conclusions. I then apply the model to particular cases in Sect. 5.4 to see how it fares against empirical scrutiny. I conclude in Sect. 5.5 with a summary of my findings and some reflections on possible policy implications.

5.2   Information Rules and Scientific Research

Information is a strange commodity with peculiar properties. As a non-rival good that costs practically nothing to duplicate, its marginal cost is essentially zero. Since welfare economics typically equates the efficient price of a good with its marginal cost of production, we might expect firms to give their informational commodities away for free. A combination of regulatory policies and basic market forces drives prices up, however. As anyone who has ever tried to found a tech company will tell you, the barriers to entry in information markets can be daunting. The obstacles to establishing and maintaining a robust technological infrastructure are numerous and expensive. Whether the product in question is hardware or software, a good or a service, high fixed costs prevent the vast majority of start-ups from succeeding in IT markets. Once a firm has captured the market, however, low marginal costs help ensure future profitability.

Successful IT firms tend to lock their consumers in through a combination of strong network effects and high switching costs (Caillaud and Jullien 2001; Klemperer 2008). The former refers to the tendency for informational goods to increase in value as more consumers use them. A classic example is the fax machine, which was practically worthless upon its introduction to the market. There is not much point in owning a fax machine if no one else has one to fax things to or from. However, as more users began buying into the technology, fax machines gradually became an essential communication tool. Similar network externalities can be observed in various IT markets, including most popular web services.

Switching costs, on the other hand, refer to the disincentive for consumers to adopt new technology once they have already become familiar with an existing system. One common example is a software suite such as Microsoft Office. Even in the face of much cheaper platforms designed to accomplish the same tasks, e.g. Open

Office, many users will choose to stick with the original program in order to avoid the hassle of learning to use the new software and converting or deleting old files. The costs associated with the loss of data and time are often more expensive than financial losses in online contexts, where competing services may be free to use.

Economists widely cite these four factors—high fixed costs and low marginal costs for producers, coupled with high switching costs and strong network effects for consumers—as the main drivers of monopolisation in information and technology markets (Shapiro and Varian 1998; Varian et al. 2004). The scientific research markets I intend to model are essentially a subset of the broader information and technology market, and thus exhibit similar features, albeit in subtly different ways. Fixed costs for cutting edge scientific research typically involve the creation of new, expensive technologies, like the Large Hadron Collider at CERN or the supercomputers currently under development for the European Commission's Human Brain Project. Once these complex machines have been built, it is relatively cheap to generate their next dataset. Thus we find that these so-called "big science" labs face the same high fixed costs and low marginal costs of production common to all producers of information and technology.

Network externalities and switching costs are also present in these markets, although scientists in this case function as both the producers and consumers of the relevant technologies. In the model developed below, the economic agents under consideration are not firms and customers, but patrons and labs. The governments, universities, and corporations that fund large-scale science projects typically do so in concert, effectively becoming a monopsony employer. The labour market is here composed of scientists, who are split into competing labs. Since there is no consumer network in this model, it is the community of scientists themselves who are subject to network externalities and switching costs as they choose whether or not to adopt new technologies. Shared databases and grid computing systems are perhaps the clearest examples of research technologies that exhibit strong network effects. With the introduction of new software platforms, it may become necessary to update or else discard data recorded in older programs, thus adding switching costs to scientific research.

A model for scientific research markets must take not just these properties into account, but also the goals of scientists, which are presumed to differ from those of classical economic agents. Rationally self-interested labs, unlike rationally self-interested firms, do not seek to maximise profits, per se. Funding is merely a means to the end of greater scientific output and attendant credit within the academic community. For the purposes of the model developed below, we will call the goods produced by such labs *discoveries*. The "revenue" from a discovery is not to be measured in financial terms, but rather in its utility for scientific progress. A "profitable" discovery is one with high utility and low technology costs. It is this modified concept of profit that labs and the monopsonist alike seek to maximise.

The emergence of monopolistic firms in information industries and monopolistic labs in scientific research markets is not necessarily the result of market failure. In some cases, monopolisation is the most efficient industrial organisation Scheme. A market in which this is the case is said to be a natural monopoly market—one that tends, through natural economic forces, to be dominated by a single agent. Utilities and other public goods are primarily produced in natural monopoly markets. This is because a competitive market for, say, gas, would result in exorbitantly expensive infrastructural redundancies. There is considerable debate as to whether information and technology markets ought to be regarded as natural monopoly industries (Wu 2010)—and, if so, what policy implications this might entail. In scientific research environments, I argue there are good reasons to presume that some—but not all—projects should be monopolised by a single lab.

With these considerations in mind, the purpose of the model proposed in Sect. 5.3 now becomes clear: to develop a modified solution to the monopsony profit maximisation problem that specifies the conditions under which patrons should endorse a natural monopoly or else foster further competition. The model draws on classical microeconomic theories of competition, monopoly, and monopsony, reframing them within an information and technology context that is specifically designed to account for the idiosyncrasies of modern scientific research.

5.3   A Model for Scientific Research Markets

As noted above, the relevant economic agents in big science research markets are labs and patrons. The former is represented by the non-empty set $L = \{L_1, L_2, \dots, L_n\}$, while the latter is a single monopsony, M. The ultimate goal of these labs is to produce discoveries $D = \{d_1, d_2, \dots d_m\}$, although to do so will require the use and/ or creation of some technology T. Discoveries may take the form of theories, inventions, or any other scientific output of value. The revenue from D is not strictly or even primarily financial, and will therefore be measured in terms of utility $U(D)$. Revenue from T will likewise be measured in terms of utility, such that $U(T)$ represents the utility of all D produced from T.

5.3.1   Monopsony Profit Maximisation

Most models of monopsony plot wages w against employment levels L, with $w(L)$ representing an increasing function of labour supply. Our factor of interest in this market, however, is technology. This variable will be plotted against costs c such that M's total production costs will be formalised as $c(T)T$. Our monopsony profit formula is therefore represented by the following equation:

$$\pi_M = U(T) - c(T)T$$

To solve M's profit maximisation problem, we take the first order condition:

$$\pi'_M = \left[U(T) - c(T)T \right]'$$

Given that M's profits are maximized at $\pi'_M = 0$, we may apply the sum rule to the latter half of the formula to reduce the equation to:

$$U'(T) = c'(T)T$$

In other words, profits are maximised when the marginal utility product of the technology (i.e., the change in D's utility from adding one more unit of T) equals its marginal cost (i.e., the change in cost from adding that unit). More formally, we may say that M's profits reach their maximum when

$$MUP_T = MC_T$$

MUP_T can also be interpreted as the demand curve for T. According to this equation, M's profits are maximised at the point of intersection between this curve and the marginal cost curve for T.

We will presume that some scientific discoveries are most efficiently achieved through competition, while others are most efficiently achieved through monopoly. When the monopsonist fails to properly identify which type of market will most efficiently produce the desired discovery, the requisite technology's marginal costs and marginal utility products are necessarily unequal. The task of this model is therefore to distinguish between these two kinds of markets.

5.3.2 Competitive vs. Natural Monopoly Markets

The true nature of the market for a discovery will depend largely upon the shape of the average cost curve for the technology required to produce it. The average cost function is defined as total costs divided by output. For our purposes, it will be plotted on a graph whose x-axis is T and y-axis is c. Note that the T-values refer specifically to discovery-bearing technology, such that one unit of T for L_1 may involve different components than the same unit of T for L_2, provided they both result in the same D. (A given computation could be made by digital or mechanical means, for instance.) By comparing the average cost curves of a lab and its competitors, it is possible to determine the most efficient market structure for a given discovery.

Say we have two AC curves, one for the monopoly lab L_M and another for the competitor lab L_C. Let us assume for simplicity's sake that both curves are linear and share the same negative slope, but that L_M's AC curve has a greater y-intercept, representing higher average costs than its competitor (see Fig. 5.1). We then mark two points on the x-axis: t_1 and t_2, such that $t_2 = 2(t_1)$. Drawing vertical lines up from these points to the two AC curves, we then draw horizontal lines across to the y-axis where t_1 hits the competitor's AC curve (c_1), and where t_2 hits the monopolist's AC curve (c_2). The area of the quadrilateral with width t_1 and height c_1 represents the

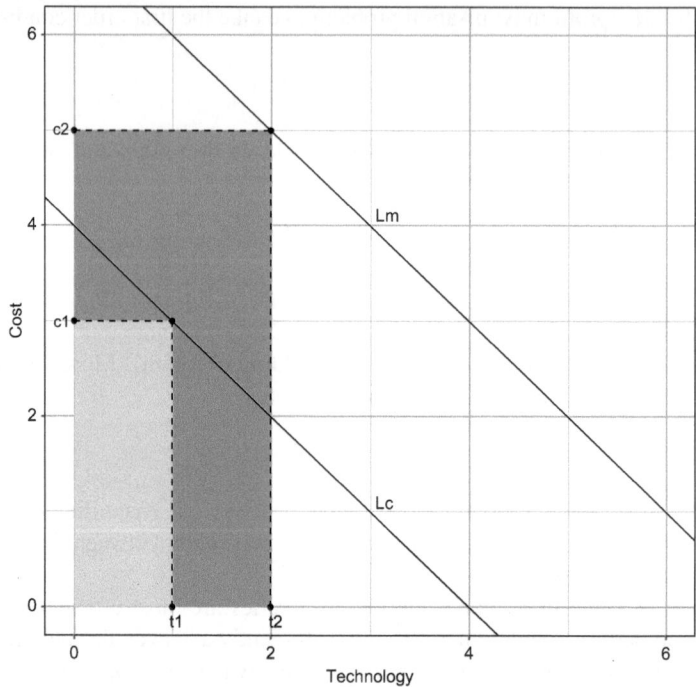

Fig. 5.1 Theoretical average cost curves for a monopoly lab (L_M) and competitor lab (L_C)

total cost for L_C to produce one unit of T. The area of the quadrilateral with width t_2 and height c_2 represents the total cost for L_M to produce two units of T. If the area of this latter quadrilateral is less than double the area of the former quadrilateral, then the market for this discovery is natural monopoly-prone.

The shaded quadrilaterals depict the total cost for each to produce one unit of technology.

The algebraic demonstration of this point is fairly straightforward. The total costs for L_C to produce one unit of T is $c_1 \times t_1$. The total cost for L_M to produce two units of T is $c_2 \times t_2$. If $(c_2 \times t_2) < 2(c_1 \times t_1)$, then the monopoly lab will produce t_2 at lower total cost to the market than two competing labs with identical AC curves, even though each on its own operates at lower average cost than the monopolist. In this case, the market will tend toward monopoly production of T. In fact, given that $t_2 = 2(t_1)$, we can reduce the inequality above to the simpler criterion: $c_2 < c_1$. If it costs an incumbent lab less to make two units of T than it costs a competing lab to make one unit of the same, then the former lab will tend to dominate the market— even if its competitors can produce each individual unit of T for less than the incumbent. Of course, depending on how established the monopoly lab is, the monopsonist may choose to cut funding for L_M altogether if other labs have demonstrated that they can do the same work for less.

This example presumes negatively sloped, linear *AC* curves, which are not necessarily the norm in IT markets. *AC* curves are more typically depicted as convex parabolas, indicating declining average costs at lower levels of output followed by higher average costs as production increases over time. The minimum point of this parabola is referred to as the minimum efficient scale (*MES*), which represents the intersection of the *AC* and *MC* curves. One common way to determine the natural structure of a market is to compare a firm's *MES* with the demand curve for the product. If the *MES* is small relative to demand, then there is room for other firms to compete in this market; if the *MES* falls near the demand curve, then this market is more likely to be dominated by a monopoly (Varian 2006).

Recall that demand in this model is identified with the technology's marginal utility product. Recall also that marginal costs in IT markets are typically small and roughly flat (Varian 2010). For the sake of simplicity, let us presume that the *MC* curve for a given *T* is a horizontal line whose *y*-intercept is $c*$. Let us further presume that the *MUP* curve is a straight line with a negative slope. Since monopsony profits are maximized when *MUP* = *MC*, the monopsonist will seek production level $t*$ such that $(t*, c*)$ represents the point of intersection between the *MUP* and *MC* curves. Then the method for differentiating between competitive and monopoly markets is essentially identical to that above.

Say that L_1 works at an *MES* of $(t_1, c*)$. If $2(t_1) > t*$, then adding a second lab L_2 with the same *MES* would be unprofitable for the monopsonist. If, however, L_2 works at an *MES* of $(t_2, c*)$, and $t_1 + t_2 < t*$, then this market has room for both labs. Generally speaking, the monopsonist should fund as many labs as possible such that the sum of the *x*-coordinates of each lab's *MES* is less than or equal to $t*$.

5.3.3   Switching Costs

The most common explanation for why certain labs tend to monopolise the market for particular discoveries is that fixed costs for big science projects are often exorbitantly high. Another compounding factor, common to all IT markets and underexplored in the literature on modern scientific research, is switching costs. For the purposes of this model, network externalities will be treated as a component of switching costs, since a lab's decision to adopt a new communication technology (e.g. a particular database infrastructure or grid computing system) will largely depend upon the number of other scientists already using it. We can broadly classify both switching costs and network externalities as two types of transaction costs (Williamson 1989).

To incorporate switching costs into this model, we will reuse the monopsony profit formula to represent a lab's profit from a given piece of technology, such that *L*'s profit from T_1 is represented by the following equation:

$$\pi_{T1} = U(T_1) - c(T_1)T_1$$

Given a new piece of technology T_2 designed to produce the same D, L must choose whether or not to adopt T_2 based on the new technology's relative profitability and the switching costs s that L would incur by updating its systems. Three outcomes are possible:

$$\pi_{T2} - \pi_{T1} > s \tag{5.1}$$

$$\pi_{T2} - \pi_{T1} = s \tag{5.2}$$

$$\pi_{T2} - \pi_{T1} < s \tag{5.3}$$

In scenario (5.1), the increase in profitability from adopting T_2 exceeds the switching costs associated with the systems update, meaning L will switch to the new technology. The opposite is true in case (5.3), implying that L will stick with T_1. In scenario (5.2), the increase in scientific profitability is equivalent to the required technological switching costs, and so L is exactly indifferent between T_1 and T_2. This case suggests a state of perfect competition between the two technologies.

Given the profit formula above, we can infer that if T_1 and T_2 are of equal utility, then L's decision whether or not to switch depends entirely upon whether the cost differential between the two technologies is greater than, equal to, or less than s. Likewise, if T_1 and T_2 are of equal cost, then L's decision whether or not to switch depends on whether the utility differential between the two technologies is greater than, equal to, or less than s. Note that if s is sufficiently large, then L will choose to stick with T_1 even if T_2 has lower costs and higher utility than T_1, provided the profit differential between the two technologies is smaller than s. If the difference on the left-hand side of the inequality in formula (5.3) is ever positive, then lock-in effects will be detrimental to scientific progress.

The analysis above is largely built upon the foundations of classic microeconomic theories. However, this model differs from more familiar accounts of monopsony and monopoly behaviour in one important respect: the T in question is not a normal commodity. As noted at the outset of Sect. 5.3, this variable refers specifically to discovery-bearing technology, which means a great deal of monopsony funding may go to waste without ever producing a single unit of T. It may take a considerable amount of time to determine whether or not a new piece of technology will lead to any discoveries, let alone the ultimate utility of those discoveries. These uncertainties and various dynamic factors confound any simple, synchronic evaluation of the relevant variables: the marginal utility product of T, its marginal cost, and, most importantly, its average cost. Accurate estimates of these functions will rely not just on economic analysis, but on the expert contributions of scientists themselves, who are in the best position to assess how close we are to making a given discovery and what technology would be required to get there.

5.4   Case Studies

The broad brushstrokes of this model now in place, it will be instructive to apply it to real world examples to see how it accords with actual funding choices. Toward that end, I turn now to two of the most prominent and ambitious big science initiatives ever undertaken: the United States Defense Department's Advanced Research Projects Agency Network (ARPANET), which would become the technological foundation of the internet; and the Human Brain Project (HBP) at the École polytechnique fédéderale de Lausanne, an ongoing international neuroscience initiative that hopes to simulate the neurochemical processes of the human brain. Both projects received funding from monopsony patrons, involve the creation and use of expensive technologies, and ultimately produced—or hope to produce—valuable discoveries.

5.4.1   Arpanet

Throughout the early- and mid-twentieth century, government and academic institutions increasingly came to rely on computers to aid in their research. The idea of establishing a communication network to connect computers in disparate locations dates back to the late 1950s, while plans to build resilient digital networks through data packet-switching protocols were originally developed in the late 1960s (Naughton 2000). ARPA began funding this research in 1969, and by 1975 the network was deemed fully operational. The transmission control and internet protocol suite (collectively known as TCP/IP) that we still use today was standardised in 1983. By 1990, the World Wide Web was running commercial services over the technology.

The government's initial investment in ARPANET was approximately $125 million over 10 years (Press 1996). This number represents the value of $c(T)T$ in the model above. The total utility of the internet is hard to fathom, but it is safe to assume it is some very large value significantly greater than ARPANET's total costs. In other words, monopsony profits from this research were very high. But the market structure for this technology changed considerably over time, and the model helps explain why. Once government researchers had a clear proposal for what they wanted ARPANET to achieve, they took bids from contractors to see who could develop the network for the cheapest price. Assuming the Defense Department had a rough idea of how much they could afford to spend on the project and that no two labs submitted bids at or under half that value, the model predicts that initial research should be conducted by a single monopoly lab. In fact, the contract was ultimately awarded to Bolt, Beranek & Newman (BBN), a technology firm who had exclusive rights to develop ARPANET at the project's outset.

Once the backbone of the network was established, researchers at other contributing institutions were invited to participate in refining the protocols. At this point, low marginal costs pulled the average cost curve, which was initially high due to large fixed costs, down toward its minimum. The two curves meet at the minimum efficient scale, which the model predicts will not be reached until a certain amount of discovery-bearing technology has already been produced. Once sceptical researchers were now beginning to realise the potential utility of packet-switching networks, and so demand for the technology, which is identified with its marginal utility product, was high. When demand is large relative to the technology's *MES*, the model predicts that more labs will become involved, which is in fact precisely what occurred in practice. From an original network of four Interface Message Processors (IMPs, functionally equivalent to modern routers), the network expanded to 13 by the end of 1970, 18 by the end of 1971, and 40 by the end of 1973.

During this period, researchers throughout the network coordinated and competed to develop better protocols. Gradually, however, TCP/IP began to emerge as the dominant protocol suite for ARPANET computers. By 1983, TCP/IP had become the official software of ARPANET, and any computer running alternative protocols would be unable to connect to the rest of the network. The switching costs of employing a different technology therefore proved prohibitively high. Network externalities ensured the dominance of a single technology, regardless of the potential profitability of others. The model predicts that TCP/IP will continue to dominate internet software until the increase in profitability from switching to an alternative protocol exceeds the switching costs associated with making the change. Given the prevalence of TCP/IP around the globe, such a switch is unlikely to happen anytime soon.

5.4.2 The Human Brain Project

Turning now from a well-known science success story to a controversial ongoing initiative, let us consider the HBP. A planned 10-year project founded in late 2013, the HBP seeks to create a complete biochemical simulation of the human brain so accurate and precise that it can effectively replace all animal and human trials for neuropharmaceuticals. Experiments could then be cheaply and safely conducted *in silico*, allowing researchers to quickly develop new drugs and bring them to market at minimal costs. If successful, the HBP would be a landmark achievement in neuroscience. It could represent a major advance in our understanding of the human brain, allowing us to better diagnose and treat a variety of neurological and psychiatric disorders that are as expensive as they are poorly understood.

Total costs for the HBP are substantial. Current estimates are that the project will cost approximately €1.2 billion over 10 years (HBP-PS Consortium 2012). More than half of that money will be spent on technology, as the HBP attempts to build a global neuroscience database designed to pool the results of all existing neuroscientific studies ever conducted. The database will be made available to researchers all over the world, so that they may upload new results and download the work of their colleagues for further analysis. Supercomputers will then aggregate the data into models that may be used to simulate neurochemical processes. Researchers hope to use these models to develop a complete simulation of the rodent brain within the next few years, followed by a complete simulation of the human brain by 2023.

Critics have alleged that these goals are implausible, pointing out that there is little precedent for such simulations in neuroscience. What work has been done in this area was far more modest in scope, simulating only small portions of the rodent brain. Moreover, there is no clear consensus in the neuroscience community as to the accuracy or utility of those simulations. In other words, there is substantial disagreement as to whether we even know what T will result in the desired D—let alone how to accurately predict the average cost or marginal utility product of such technologies. This uncertainty complicates matters for the monopsonist.

The preliminary task of pooling the world's neuroscientific data into one database, while certainly ambitious, would seem an obvious first step toward any future breakthroughs in the field. The task of determining what technologies will best utilise that data can only begin once the platform itself is created. Large databases tend to be monopolised, since the whole point is to consolidate information in one place. Unless two competing labs can each accomplish this task for less than half the cost of the HBP, then the model predicts that it will be most efficient to let researchers monopolise the market for this technology.

Once the database is created, however, the calculus changes. Active disagreement in the scientific community would seem to be *prima facie* evidence that the market for this discovery is not yet ready for a major monopsony investment. Since the European Union is apparently willing to spend large sums of money on this research, however, we may conclude that demand for this D is quite high. With a high, fixed MUP and uncertainty as to the value of the technology's AC or MC functions, the most cautiously efficient use of funds would be to split monopsony patronage amongst a series of competing labs with more modest discovery goals. If and when one lab's technology produces far more discoveries than those of its competitors, the monopsonist will redirect investment toward the most successful lab in order to capitalise on its newfound profitability.

While HBP research is still in its early stages, it appears that the EU is sensitive to the potential inefficiencies of their patronage. The European Commission's 2015 report on the HBP's progress urged researchers to clarify their goals, avoid unrealistic expectations, and improve coordination between subprojects (European Commission 2015). Their recommendations, while not precisely identical to those proposed above, will modify the HBP's mission and strategy to make them more consonant with those predicted by the model.

These two case studies demonstrate that the model for scientific research markets developed in Sect. 5.3 is applicable across a variety of subjects, times, and places. In both instances—and in modern scientific research more generally—the role of information and communication technology in cutting edge research has major economic impacts on the proper funding strategy for the desired discovery.

5.5   Conclusion

Funding for scientific research is an essential part of any advanced nation's budget. The most important and ambitious projects across the sciences increasingly rely on expensive, new technologies. To help foster further scientific progress, the governments and universities that fund such research must pay close attention to the economic principles governing industry structure for technology markets. While these principles have been extensively explored in commercial contexts, their significance for scientific research markets is not yet widely appreciated. Accurately predicting the ideal market for a given scientific discovery requires thoughtful evaluation of the marginal utility product, marginal cost, and average cost functions for the technology required to make it. Improper estimates of these values will result in either monopoly control of markets that should be competitive, or competition in natural monopoly markets. Either scenario would come at the detriment of monopsony profits.

Of course, the scientists who already have the ear of their monopsony patrons will likely use their influence to promote their own research. In this way, the feedback effect of power in scientific research markets resembles monopoly behaviour in other information and technology industries, where incumbents are in a privileged position to dictate the terms of the market and even design their own regulatory regimes. To confirm the efficiency of their funding, the monopsonist will need to commission frequent assessments of projected and actual values for the *MUP*, *MC*, and *AC* functions of the technology used in pursuit of a given *D*. Moreover, historical analysis will be necessary to determine whether or not the use of promising technologies was unduly discouraged by high switching costs and/or network externalities. These economic considerations are a vital part of the effort to ensure ongoing scientific progress.

Acknowledgements   I would like to thank Greg Taylor and Carl Öhman for their helpful comments on an earlier draft of this chapter.

References

Caillaud, B., and B. Jullien. 2001. Competing cybermediaries. *European Economic Review* 45 (4–6): 797–808.

European Commision. 2015. *The human brain project flagship: First technical project review.* Retrieved 25 April, 2015 from: http://ec.europa.eu/digital-agenda/en/news/1st-technical-review-human-brain-project-hbp-main-conclusions-recommendations

HBP-PS Consortium. 2012. *The human brain project: A report to the European Commission.* Retrieved 25 April, 2015 from: https://www.humanbrainproject.eu/documents/10180/17648/TheHBPReport_LR.pdf

Klemperer, P. 2008. Network effects and switching costs. In *The new Palgrave dictionary of economics*, ed. S.N. Durlauf and L.E. Blume, 2nd ed. New York: Palgrave Macmillan.

Naughton, J. 2000. *A brief history of the future*. London: Orion Books.

Press, L. 1996. Seeding networks: The federal role. *Communications of the ACM* 39 (10): 11–18.

Shapiro, C., and H. Varian. 1998. *Information rules: A strategic guide to the network economy.* Cambridge, MA: Harvard Business School Press.

Varian, H. 2006. *Intermediate microeconomics: A modern approach.* 8th ed. New York: W.W. Norton.

Varian, H., J. Farrell, and C. Shapiro. 2004. *The economics of information technology: An introduction.* Cambridge: Cambridge University Press.

Wu, T. 2010. *The master switch.* New York: Alfred A. Knopf.

Chapter 6
Projecting AI-Crime: A Review of Plausible Threats

Thomas King

Abstract Artificial Intelligence (AI) research and regulation seek to balance the benefits of innovation against any potential harms and disruption. However, one unintended consequence of the recent surge in AI research is the potential re-orientation of AI technologies to facilitate criminal acts, which I term AI-Crime (AIC). We already know that AIC is theoretically feasible thanks to published experiments in automating fraud targeted at social media users, as well as demonstrations of AI-driven manipulation of simulated markets. However, because AIC is still a relatively young and inherently interdisciplinary area of research—spanning socio-legal studies to formal science—there is little certainty of what an AIC future might look like. This article offers the first systematic, interdisciplinary literature analysis of the foreseeable threats of AIC, providing law enforcement and policy-makers with a synthesis of the current problems.

6.1 Introduction

Artificial Intelligence (AI) may play an increasingly essential[1] role in criminal acts in the future. Criminal acts are defined here as any act (or ommission) constituting an offence punishable under English criminal law, without loss of generality to jurisdictions that similarly define crime. Evidence of what I shall call "AI-Crime" (AIC) is provided by two (theoretical) research experiments. In the first one, two computational social scientists (Seymour and Tully 2016) used AI as an instrument

---

[1] I use "essential" (in place of "necessary") to indicate that while there is a logical possibility that the crime could occur without the support of AI, this possibility is negligible. That is, the crime would probably not have occurred but for the use of AI. The distinciton can be clarified with an example. One might consider transport to be *essential* to travel between Paris and Rome, but one could always walk: transport is not in this case (strictly speaking), *necessary*. Furthermore, note that AI-crimes as defined in this article involve AI as a contributory factor, but not an investigative, enforcing, or mitigating factor.

T. King (✉)
Oxford Internet Institute, Digital Ethics Lab, University of Oxford, Oxford, UK
e-mail: thomasc.king@oii.ox.ac.uk

© Springer Nature Switzerland AG 2019   65
C. Öhman, D. Watson (eds.), *The 2018 Yearbook of the Digital Ethics Lab*,
Digital Ethics Lab Yearbook, https://doi.org/10.1007/978-3-030-17152-0_6

to convince social media users to click on phishing links within mass-produced messages. Because each message was constructed using machine learning techniques applied to users' past behaviours and public profiles, the content was tailored to each individual, thus camouflaging the intention behind each message. If the potential victim had clicked on the phishing link and filled in the subsequent web-form, then (in real-world circumstances) a criminal would have obtained personal and private information that could be used for theft and fraud. AI-fuelled crime may also impact commerce. In the second experiment, three computer scientists (Martínez-Miranda et al. 2016) simulated a market and found that trading agents could learn and execute a "profitable" market manipulation campaign comprising a set of deceitful false-orders. These two experiments show that AI provides a feasible and fundamentally novel threat, in the form of AIC.

The importance of AIC as a distinct phenomenon has not yet been acknowledged. The literature on AI's ethical and social implications focuses on regulating and controlling AI's civil uses, rather than considering its possible role in crime (Kerr 2004). Furthermore, the AIC research that is available is scattered across disciplines, including socio-legal studies, computer science, psychology, and robotics, to name just a few. This lack of research centred on AIC undermines the scope for both projections and solutions in this new area of potential criminal activity.

To provide some clarity about current knowledge and understanding of AIC, this chapter offers an analysis of the relevant, interdisciplinary academic literature, in order to give an up to date answer to the following question:

1. What are the fundamentally unique but plausible threats surrounding AIC?

To address this question I analyse: the potential areas of AIC according to the literature, and the more general concerns that cut across them.

Given that the aforemention question must be answered to support a normative foresight analysis, the analysis here focuses only on *realistic* and *plausible* concerns surrounding AIC. We should not be interested in speculations unsupported by scientific knowledge or empirical evidence. Consequently, the analysis is based on the classical definition of AI provided by McCarthy, Minsky, Rochester, and Shannon in their seminal "Proposal for the Dartmouth Summer Research Project on Artificial Intelligence", the founding document and later event that established the new field of AI in 1955:

> For the present purpose the artificial intelligence problem is taken to be that of making a machine behave in ways that would be called intelligent if a human were so behaving. (McCarthy et al. 1955)

As (Floridi 2017) argues, this is a counterfactual: were a human to behave in that way we would call that behaviour intelligent. It does not mean that the machine *is* intelligent. The latter scenario is a fallacy, and smacks of superstition. The same understanding of AI underpins the Turing test (Floridi et al. 2009), which checks the ability of a machine to perform a task in such a way that the *outcome* would be indistinguishable from the outcome of a human agent working to achieve the same task (Turing 1950). In other words, I define AI based on outcomes and actions.

This definition identifies in AI applications a growing resource of interactive, autonomous, and self-learning *agency*, to deal with tasks that would otherwise require human intelligence and intervention to be performed successfully. Such artificial agents (AAs) as noted by (Floridi and Sanders 2004, p. 3) are "sufficiently informed, 'smart', autonomous and able to perform morally relevant actions independently of the humans who created them [...] (James Gips 1995)."

This combination of autonomy and learning skills underpins (as discussed by Yang et al. 2018) both beneficial and malicious uses of AI.[2] I shall therefore treat AI in terms of a *reservoir of smart agency on tap*.

6.2   General Threat Categories

The four general threat categories discussed here capture what makes AIC possible compared to crimes of the past (e.g., the reasons behind AI's potential dual-use), and uniquely problematic (this is the justification for conceptualising AIC as a distinct crime-phenomenon). These categories are emergence, liability, monitoring, and psychology.

Emergence of artificial agent behaviour is a problem whereby potentially criminal acts or omissions unexpectedly evolve out of the relatively simple design and implementation of an artificial agent (AA). For example from machine learning techniques applied to the ordinary interaction between agents in a multi-agent system (MAS), coordinated actions and plans may emerge autonomously. Such emergence of complexity may, in some cases, be promoted by a designer as a property that ensures that specific solutions are discovered at run-time based on general goals issued at design-time. Such relatively simple design leading to more complex behaviour is a core desideratum of MASs (Hildebrandt 2008, p.7). In other cases, a designer may want to prevent emergence, such as when an autonomous trading agent inadvertently coordinates and colludes with other trading agents towards a shared goal (Martínez-Miranda et al. 2016). I can see then, that emergent behaviour may have criminal implications, insofar as it misaligns with the original design. As (Alaieri and Vellino 2016, p. 1) put it, "non-predictability and autonomy may confer a greater degree of responsibility to the machine but it also makes them harder to trust".

Liability for an AI-crime becomes problematic under existing liability doctrines, thereby threatening the dissuasive and redressing power of the law. Existing liability models may be inadequate to address the future role of AI in criminal activities. The limits of the liability models may therefore undermine the certainty of the law, as it may be the case that agents, artificial or otherwise, may perform criminal acts or omissions without sufficient concurrence with the conditions of liability for a

---

[2] Because much of AI is fueled by data, some of its challenges are rooted in data governance (Cath et al. 2017), particularly issues of consent, discrimination, fairness, ownership, privacy, surveillance, and trust (Floridi and Taddeo 2016).

particular offence to constitute a (specifically) criminal offence. The first condition of criminal liability is the *actus reus*: a voluntarily taken criminal act or omission. For types of AIC defined such that only the AA can carry out the criminal act or omission, the voluntary aspect of *actus reus* may never be met since the idea that an AA can act voluntarily is contentious:

> the conduct proscribed by a certain crime must be done voluntarily. What this actually means it is something yet to achieve consensus, as concepts as consciousness, will, voluntariness and control are often bungled and lost between arguments of philosophy, psychology and neurology (Freitas et al. 2014, p. 9).

When criminal liability is fault-based, it also has a second condition, the *mens rea* (a guilty mind), of which there are many different types and thresholds of mental state applied to different crimes. In the context of AIC, the *mens rea* may comprise an intention to commit the *actus reus* using an AI-based application (intention threshold) or knowledge that deploying an AA will or could cause it to perform a criminal action or omission (knowledge threshold).

Concerning an intention threshold, if we admit that an AA can perform the *actus reus*, in those types of AIC where intention (partly) constitutes the *mens rea*, greater AA autonomy increases the chance of the criminal act or omission being decoupled from the mental state (intention to commit the act or omission) "autonomous robots [and AAs] have a unique capacity to splinter a criminal act, where a human manifests the mens rea and the robot [or AA] commits the actus reus" (McAllister 2017, p. 47).

Concerning the knowledge threshold, in some cases the *mens rea* could actually be missing entirely. The potential absence of a knowledge-based *mens rea* is due to the fact that, even if we understand that an AA can perform the *actus reus* autonomously, the complexity of the AA's programming makes it possible that the designer, developer, or deployer (i.e., a human agent) will neither know nor predict the AA's criminal act or omission. The implication is that the complexity of AI

> provides a great incentive for human agents to avoid finding out what precisely the ML [machine learning] system is doing, since the less the human agents know, the more they will be able to deny liability for both these reasons (Williams 2017).

Alternatively, legislators may define criminal liability without a fault requirement. Such faultless liability, which is increasingly used for product liability tort law (e.g., pharmaceuticals and consumer goods) would lead to liability being assigned to the faultless legal person who deployed an AA despite the risk that it may conceivably perform a criminal action or omission. Such faultless acts may involve many human agents contributing to the *prima facie* crime, such as through programming or deployment of an AA. Determining who is responsible may therefore rest with the faultless responsibility approach for distributed moral actions (Floridi 2016). In this distributed setting, liability is applied to the agents who *make a difference* in a complex system in which individual agents perform neutral actions that nevertheless result in a collective criminal one. However, some argue (Williams 2017) that *mens rea* with intent or knowledge is central to the criminal law's entitlement to censure (Ashworth 2010) and we cannot simply abandon that key requirement [a common key requirement] of criminal liability in the face of difficulty in proving it (Williams 2017).

The problem is that, if *mens rea* is not entirely abandoned and the threshold is only lowered, then for balancing reasons the punishment may be too light (the victim is not adequately compensated) and yet simultaneously disproportionate (was it really the defendant's fault?) in the case of serious offences, such as those against the person (McAllister 2017, p. 38).

Monitoring AIC faces three kinds of problem: attribution, feasibility, and cross-system actions. Attributing non-compliance with current legislation is a problem of monitoring of AAs used as instruments of crime, due to the capacity of this new type of smart agency to act independently and autonomously, two features that will muddle any attempt to trace an accountability trail back to a perpetrator.

Concerning the feasibility of monitoring, a perpetrator may take advantage of cases where AAs operate at speeds and levels of complexity that are simply beyond the capacity of compliance monitors. An exemplary case is of AAs that integrate into mixed human and artificial systems in ways that are hard to detect, such as social media bots. Social media sites can hire experts to identify and ban malicious bots (for example, no social media bot is currently capable of passing the Turing test (Wang et al. 2012).[3] Nonetheless, because deploying bots is far cheaper than employing people to test and identify each bot, the defenders (social media sites) are easily outscaled by the attackers (criminals) that deploy the bots (Ferrara et al. 2014, p. 5). Detecting bots at low cost is possible by using machine learning as an automated discriminator, as suggested by (Ratkiewicz et al. 2011). However, a discriminator is both trained and claimed as effective using data comprising known bots, which may be substantially less sophisticated than more evasive bots used by malevolent actors, which may therefore go undetected in the environment (Ferrara et al. 2014). Such potentially sophisticated bots may also use machine-learning tactics in order to adopt human traits, such as posting according to realistic circadian rhythms (Golder and Macy 2011), thus evading machine-learning based detection. All of this may obviously lead to an arms race in which attackers and defenders mutually adapt to each other (Alvisi et al. 2013; Zhou and Kapoor 2011, p. 1)—presenting a serious problem in an offence-persistent environment such as cyberspace (Seymour and Tully 2016; Taddeo 2017).

Cross-system actions pose a problem for tunnel-visioned AIC monitors that only focus on a single system. Cross-system experiments (Bilge et al. 2009) show that automated copying of a user's identity from one social network to another (a cross-system identity theft offence) is more effective at deceiving other users than copying an identity from within that network. In this case, the social network's policy may be at fault. Twitter, for example, takes a rather passive role, only banning cloned profiles when users submit reports, rather than by undertaking cross-site validation

---

[3] Claims to the contrary can be dismissed as mere hype, the result of specific, *ad hoc* constraints, or just tricks, see for exmaple the chatterbot named "Eugene Goostman", see https://en.wikipedia.org/wiki/Eugene_Goostman

(Twitter – Impersonation policy 2018). *Psychology* encapsulates the threat of AI that affects a user's mental state to the (partial or full) extent of facilitating or causing crime. One psychological effect rests on the capacity for AAs to gain trust from users, making them vulnerable to manipulation, as (Weizenbaum 1976) already indicated a long time ago, after conducting early experiments into human-bot interaction where people revealed unexpectedly personal details about their lives.

Psychology encapsulates the threat of AI affecting a user's mental state to the (partial or full) extent of facilitating or causing crime. One psychological effect rests on the capacity for AAs to gain trust from users, making people vulnerable to manipulation. This was demonstrated some time ago by (Weizenbaum 1976), after conducting early experiments into human–bot interaction where people revealed unexpectedly personal details about their lives.

We are now ready to understand the specific areas of AIC and the involvement of these more general threats in detail.

6.3 Crime Areas

Naturally, there are many types of crime we could consider as a type of AIC. Here, I narrow down the selection to just those identified in the literature that meet the following conditions:

- have occurred or will likely occur according to current AI technologies (*plausibility*);
- require AI as an essential factor (*uniqueness*)[4]; and
- are criminalised in domestic law (i.e., I exclude international crimes, e.g., war-related).

These criteria allow us to extract the areas of crime defined by the UK's leading criminal law practitioner's textbook (Archbold 2018) that are potential areas of AIC. There are four such AIC areas in total, summarised in Table 6.1, each of which falls under at least one general threat category.

Table 6.1 A map of areas-specific and cross-cutting threats

	Emergence	Liability	Monitoring	Psychology
Commerce, financial markets, and insolvency	✓	✓	✓	
Harmful or dangerous drugs			✓	✓
Offences against the person	✓	✓		
Theft and fraud, and forgery and personation			✓	

---

[4] However, I did not require AI's role to be sufficient for the crime because normally other technical and non-technical elements are likely to be needed. For example, if robotics are instrumental (e.g., involving autonomous vehicles) or causal in crime, then any underlying AI component must be essential for the crime to be included in my analysis.

6.3.1 Commerce, Financial Markets, and Insolvency

This economy-focused area of crime is defined in (Archbold 2018, Ch. 30) and includes *cartel offences*, such as price fixing and collusion, *insider dealing*, such as trading securities based on private business information, and *market manipulation*. The literature that I analysed raises concerns over AI's involvement in market manipulation, price fixing, and collusion.

Market manipulation is defined as "actions and/or trades by market participants that attempt to influence market pricing artificially" (Spatt 2014, p. 1), where a necessary criterion is an intention to deceive (Wellman and Rajan 2017, p. 11). Yet, such deceptions have been shown to emerge from a seemingly compliant implementation of an AA that is designed to trade on behalf of a user (that is, an artificial trading agent). This is because an AA, "particularly one learning from real or simulated observations, may learn to generate signals that effectively mislead" (Wellman and Rajan 2017, p. 14). Simulation-based models of markets comprising artificial trading agents have shown (Martínez-Miranda et al. 2016) that, through reinforcement learning, an AA can learn the technique of order-book spoofing.

This involves "placing orders with no intention of ever executing them and merely to manipulate honest participants in the marketplace" (Lin et al. 2017, p. 37). In this case, the market manipulation emerged from an AA initially exploring the action space and, through exploration, placing false orders that became *reinforced* as a profitable strategy, and subsequently exploited for profit (Martínez-Miranda et al. 2016). Further market exploitations, this time involving human intent, also include

> acquiring a position in a financial instrument, like a stock, then artificially inflating the stock through fraudulent promotion before selling its position to unsuspecting parties at the inflated price, which often crashes after the sale (Lin et al. 2017, p. 32).

This is colloquially known as a pump-and-dump scheme. Social bots have been shown to be effective instruments of such schemes. For instance, in a recent prominent case a social bot network's sphere of influence was used to spread disinformation about a barely traded public company. The company's value gained "more than 36,000% when its penny stocks surged from less than \$0.10 to above \$20 a share in a matter of few weeks" (Ferrara 2015, p. 2). Although such social media spam is unlikely to sway most human traders, algorithmic trading agents act precisely on such social media sentiment (Sætenes 2017, p. 3). These automated actions can have significant effects for low-valued (under a penny) and illiquid stocks, which are susceptible to volatile price swings (Lin et al. 2017).

Collusion, in the form of price fixing, may also emerge in automated systems thanks to AAs' planning and autonomy capabilities. Empirical research finds two necessary conditions for (non-artificial) collusion:

> (1) those conditions which lower the difficulty of achieving effective collusion by making coordination easier; and (2) those conditions which raise the cost of non-collusive conduct by increasing the potential instability of non-collusive behavior (Hay and Kelley 1974, p. 3).

Near-instantaneous pricing information (e.g., via a computer interface) meets the coordination condition. When agents develop price-altering algorithms, any action to lower a price by one agent may be instantaneously matched by another. In and of itself, this is no bad thing and only represents an efficient market. Yet, the possibility that lowering a price will be responded in kind is discincentivising and hence meets the punishment condition. Therefore, if the shared strategy of price-matching is common knowledge,[5] then the algorithms (if they are rational) will maintain artificially and tacitly agreed higher prices, by not lowering prices in the first place (Ezrachi and Stuck 2016, p. 5). Crucially, for collusion to take place, an algorithm does not need to be designed specifically to collude. As Ezrachi and Stucke argue, "artificial intelligence plays an increasing role in decision making; algorithms, through trial-and-error, can arrive at that outcome [collusion]" (Ezrachi and Stuck 2016, p. 5). The lack of intentionality, the very short decision span, and the likelihood that collusion may emerge as a result of interactions among AAs also raises serious problems with respect to liability and monitoring. Problems with liability refer to the possibility that

> the critical entity of an alleged [manipulation] scheme is an autonomous, algorithmic program that uses artificial intelligence with little to no human input after initial installation (Lin et al. 2017, p. 49).

In turn, the autonomy of an AA raises the question as to whether

> regulators need to determine whether the action was intended by the agent to have manipulative effects, or whether the programmer intended the agent to take such actions for such purposes? (Wellman and Rajan 2017, p. 4).

Monitoring becomes difficult in the case of financial crime involving AI, because of the speed and adaptation of AAs. High-speed trading

> encourages further use of algorithms to be able to make automatic decisions quickly, to be able to place and execute orders and to be able to monitor the orders after they have been placed (Lier 2016, p. 39).

Artificial trading agents adapt and "alter our perception of the financial markets as a result of these changes" (Lier 2016, p. 37). At the same time, the ability of AAs to learn and refine their capabilities implies that these agents may evolve new strategies, making it increasingly difficult to detect their actions (Farmer and Skouras 2013). Moreover, the problem of monitoring is inherently one of monitoring a system-of-systems, because the capacity to detect market manipulation is affected by the fact that its effects

> in one or more of the constituents may be contained, or may ripple out in a domino-effect chain reaction, analogous to the crowd-psychology of contagion (Cliff and Northrop 2012, p. 12).

---

[5] Common knowledge is a property found in epistemic logic about a proposition P and a set of agents. P is common knowledge if and only if each agent knows P, each agent knows the other agents know P, and so on. Agents may acquire common knowledge through broadcasts, which provide agents with a rational basis to act in coordination (e.g., collectively turning up to a meeting following the broadcast of the meeting's time and place).

Cross-system monitoring threats may emerge if and when trading agents are deployed with broader actions, operating at a higher level of autonomy across systems, such as by reading from or posting on social media (Wellman and Rajan 2017). These agents may, for example, learn how to engineer pump-and-dump schemes, which would be unseeable from a single-system perspective.

6.3.2 Harmful or Dangerous Drugs

Crimes falling under this category include *trafficking*, *selling*, *buying*, and *possessing banned drugs* (Archbold 2018, Ch. 27). The literature I surveyed finds that AI can be instrumental in supporting the trafficking and sale of banned substances.

The literature raises the business-to-business trafficking of drugs as a threat due to criminals using unmanned vehicles, which rely on AI planning and autonomous navigation technologies, as instruments for improving success rates of smuggling. Because smuggling networks are disrupted by monitoring and intercepting transport lines, law enforcement becomes more difficult when unmanned vehicles are used to transport contraband. Drones present, according to Europol (Europol 2017), a horizonal threat in the form of automated drug smuggling. Remote-controlled cocaine-trafficking submarines have already been discovered and seized by US law enforcement (Sharkey et al. 2010).

Unmanned underwater vehicles (UUVs) offer a good example of the dual-use risks of AI, and hence of the potential for AIC. UUVs have been developed for legitimate uses (e.g., defence, border protection, water patrolling) and yet they have also proven effective for illegal activities, posing, for example, a significant threat to enforcing drug prohibitions. Criminals can, presumably, avoid implication because UUVs can act independently of an operator (Gogarty and Hagger 2008, p. 3). Hence, no link with the deployer of the UUVs can be ascertained positively, if the software (and hardware) lacks a breadcrumb trail back to who obtained it and when, or if the evidence can be destroyed upon the UUV's interception (Sharkey et al. 2010). Controlling the manufacture of submarines and hence traceability is not unheard of, as reports on the discovery in the Colombian coastal jungle of multi-million dollar manned submarines illustrate. However, such manned submarines risk attribution to the crew and the smugglers, unlike UUVs. In Tampa, Florida, over 500 cases were successfully brought against smugglers using manned submarines between 2000–2016, resulting in an average 10-year sentence (Marrero 2016). Hence, UUVs present a distinct advantage compared to traditional smuggling approaches.

The literature is also concerned with the drugs trade's business-to-consumer side. Already, machine learning algorithms have detected advertisements for opioids sold without prescription on Twitter (Mackey et al. 2017). Because social bots can be used to advertise and sell products, (Kerr and Bornfreund 2005, p. 8) ask whether these buddy bots [that is, social bots] could be programmed to send and reply to email or use instant messaging (IM) to spark one-on-one conversations with hundreds of thousand or even millions of people every day, offering pornography or *drugs* to children, *preying on teens' inherent insecurities to sell them needless products and services* (emphasis mine).

As the authors outline, the risk is that social bots could exploit cost-effective scaling of conversational and one-to-one advertising tools to facilitate the sale of illegal drugs.

6.3.3   Offences Against the Person

Crimes that fall under offences against the person range from murder to human trafficking (Archbold 2018, Chapter 19), but the literature that the analysis uncovered exclusively relates AIC to *harassment* and *torture*. Harassment comprises intentional and repetitious behaviour that alarms or causes a person distress. Harassment is, according to past cases, constituted by at least two incidents or more against an individual (Archbold 2018, secs. 19–354). Regarding torture, (Archbold 2018, secs. 19–435) states that:

a public official or person acting in an official capacity, whatever his nationality, commits the offence of torture if in the United Kingdom or elsewhere he intentionally inflicts severe pain or suffering on another in the performance or purported performance of his official duties.

Concerning harassment-based AIC, the literature implicates social bots. A malevolent actor can deploy a social bot as an instrument of direct and indirect harassment. Direct harassment is constituted by spreading hateful messages against the person (Mckelvey and Dubois 2017, p. 16). Indirect methods include retweeting or liking negative tweets and skewing polls to give a false impression of wide-scale animosity against a person (Mckelvey and Dubois 2017, p. 16). Additionally, a potential criminal can also subvert another actor's social bot, by skewing its learned classification and generation data structures via user-interaction (i.e., conversation). This is what happened in the case of Microsoft's ill-fated social Twitter bot "Tay", which quickly learned from user-interactions to direct "obscene and inflammatory tweets" at a feminist-activist (Neff and Nagy 2016). Because such instances of what might be deemed harassment can become entangled with the use of social bots to exercise free speech, jurisprudence must demarcate between the two to resolve ambiguity (Mckelvey and Dubois 2017, p. 16). Some of these activities may comprise harassment in the sense of socially but not legally unacceptable behaviour, whilst other activities may meet a threshold for criminal harassment.

Liability also proves to be problematic in these cases. In the case of Tay, critics "derided the decision to release Tay on Twitter, a platform with highly visible problems of harassment" (Neff and Nagy 2016, p. 8). Yet, as Neff and Nagy go on to say, users are also to be blamed if "technologies should be used properly and as they were designed". Differing perspectives and opinions on harassment by social bots are inevitable in such cases where the *mens rea* of a crime is considered (strictly) in terms of intention, because attribution of intent is a non-agreed function of at least engineering, application context, human-computer interaction, and perception.

Concerning torture, the AIC risk becomes plausible if and when developers integrate AI planning and autonomy capabilities into an interrogation AA. This is the case with automated detection of deception in a prototype robotic guard for the United States' border control (Nunamaker Jr. et al. 2011). Using AI for interrogation is motivated by its capacity for better detection of deception, human trait emulation (e.g., voice), and affect-modelling to manipulate the interrogatee (McAllister 2017, p. 17). Yet, an AA with these capabilities may learn to torture a victim (McAllister 2017 p. 39). For the interrogation subject, the risk is that an AA may be deployed to apply psychological (e.g., mimicking people known to the torture subject) or physical torture techniques. Despite misconceptions, experienced professionals report that torture (in general) is an ineffective method of information extraction (Janoff-Bulman 2007). Nevertheless, some malicious actors may perceive the use of AI as a way to optimise the balance between suffering, and causing the interogatee to lie, or become confused or unresponsive. All of this may happen independently of human intervention.

Such distancing of the perpetrator from the *actus reus* is another reason torture falls under AIC as a unique threat, with three factors that may particularly motivate the use of AAs for torture (McAllister 2017, pp. 19–20). First, the interrogatee likely knows that the AA cannot understand pain or experience empathy, and is therefore unlikely to act with mercy and stop the interrogation. Without compassion the mere presence of an interrogation AA may cause the subject to capitulate out of fear, which, according to international law, is possibly but ambiguously a crime of (threatening) torture (Solis 2016, pp. 437–485). Second, the AA's deployer may be able to detach themselves emotionally. Third, the deployer can also detach themselves physically (i.e., will not be performing the *actus reus* under current definitions of torture). It therefore becomes easier to use torture, as a result of improvements in efficacy (lack of compassion), deployer motivation (less emotion), and obfuscated liability (physical detachment). Similar factors may entice state or private corporations to use AAs for interrogation. However, banning AI for interrogation (McAllister 2016, p. 5) may face a push back similar to the one seen with regard to banning autonomous weapons. "Many consider [banning] to be an unsustainable or impractical solution (Anderson and Waxman 2013; Solis 2016)" if AI offers a perceived benefit to overall protection and safety of a population, making limitations on use rather than a ban a potentially more likely option.

Liability is a pressing problem in the context of AI-driven torture (McAllister 2017, p. 24). As for any other form of AIC, an AA cannot itself meet the *mens rea* requirement. Simply, an AA does not have any intentionality, nor does it have the ability to ascribe meaning to its actions. Indeed, an argument that applies to the current state-of-the-art (and perhaps beyond) is that computers (which implement AAs) are syntactic, not semantic, machines (Searle 1983), meaning that they can perform actions and manipulations but without ascribing any meaning to them: any meaning is situated purely in the human operators (Taddeo and Floridi 2005). As unthinking machines, AAs therefore cannot bear moral responsibility or liability for their actions. However, taking an approach of strict criminal liability, where punishment or damages may be imposed without proof of fault, may offer a way out of the problem by lowering the intention-threshold for the crime, rather than relying solely on civil liability.

In this respect, (McAllister 2017, p. 38) argues that strict civil liability is inappropriate due to the unreasonable degree of foresight required of a designer when an AA learns to torture in unpredictable ways. Developing and deploying such unpredictable and complex autonomous AI-interrogators means, as Grut stresses, that

> it is even less realistic to expect human operators [or deployers] to exercise significant veto control over their operations (Grut 2013, p. 11).

Even if control is not an issue, the typical punishment for product fault liability is a fine, which may be neither equitable nor dissuasive given the potential seriousness of any ensuing human rights violations (McAllister 2017, p. 38). Hence, serious strict-liability AIC would require specific sentencing guidelines to be developed to impose punishments fitting the offence, even if the offence is not intentional, as is the case with corporate manslaughter, where individuals can be given prison sentences. The question of who exactly should face imprisonment for AI-caused offences against the person (as for many uses of AI), is difficult and is significantly hampered by the 'problem of many hands' (van de Poel et al. 2012). It is clear that an AA cannot be held liable. Yet, the multiplicity of actors creates a problem in ascertaining where the liability lies—whether with the person who commissioned and operated the AA, or its developers, or the legislators and policymakers who sanctioned (or didn't prohibit) real-world deployment of such agents (McAllister 2017, p. 39). Serious crimes (including both physical and mental harm) that have not been foreseen by legislators might plausibly fall under AIC, with all the associated ambiguity and lack of legal clarity. This motivates the extension or clarification of existing joint liability doctrines.

6.3.4   Theft and Fraud, and Forgery and Personation

The literature I reviewed connects forgery and personation via AIC to theft and non-corporate fraud, and also implicates the use of machine learning in corporate fraud.

Concerning theft and non-corporate fraud, the literature describes a two-phase process that begins with using AI to gather personal data and proceeds to using stolen personal data and other AI methods to forge an identity that convinces the banking authorities to make a transaction (that is, involving banking theft and fraud). In the first phase of the AIC pipeline for theft and fraud, there are three ways for AI techniques to assist in gathering personal data.

The first method involves using social media bots to target users at large scale and low cost, by taking advantage of their capacity to generate posts, mimic people, and subsequently gain trust through friendship requests or "follows" on sites like

Twitter, LinkedIn, and Facebook (Bilge et al. 2009). When a user accepts a friendship request, a potential criminal gains personal information, such as the user's location, telephone number, or relationship history, which are normally only available to that user's accepted friends (Bilge et al. 2009). Because many users add so-called friends whom they do not know, including bots, such privacy-compromising attacks have an unsurprisingly high success rate. Past experiments with a social bot exploited 30–40% of users in general (Bilge et al. 2009) and 60% of users who shared a mutual friend with the bot (Boshmaf et al. 2012). Moreover, identity-cloning bots have succeeded, on average, in having 56% of their friendship requests accepted on LinkedIn (Bilge et al. 2009). Such identity cloning may raise suspicion due to a user appearing to have multiple accounts on the same site (one real and one forged by a third-party). Hence, cloning an identity from one social network to another circumvents these suspicions, and in the face of inadequate monitoring such cross-site identity cloning is an effective tactic (Bilge et al. 2009), as discussed above.

The second method to gather personal data, which is compatible with and may even build on the trust gained via friending social media users, makes partial use of conversational social bots for social engineering (Alazab and Broadhurst 2016, p. 12). This occurs when AI "attempts to manipulate behaviour by building rapport with a victim, then exploiting that emerging relationship to obtain information from or access to their computer (Chantler and Broadhurst 2006)".

Although the literature seems to support the efficacy of such bot-based social-engineering, given the currently limited capabilities of conversational AI, scepticism is justified when it comes to automated manipulation on an individual and long-term basis. However, as a short-term solution, a criminal may cast a deceptive social botnet sufficiently widely to discover susceptible individuals. Initial AI-based manipulation may gather harvested personal data and re-use it to produce "more intense cases of simulated familiarity, empathy, and intimacy, leading to greater data revelations" (Graeff 2014). After gaining initial trust, familiarity and personal data from a user, the (human) criminal may move the conversation to another context, such as private messaging, where the user assumes that privacy norms are upheld (Graeff 2014). Crucially, from here, overcoming the conversational deficiencies of AI to engage with the user is feasible using a cyborg; that is, a bot-assisted human (or vice versa) (Chu et al. 2010). Hence, a criminal may make judicious use of the otherwise limited conversational capabilities of AI as a plausible means to gather personal data.

The third method for gathering personal data from users is automated phishing. Ordinarily, phishing is unsuccessful if the criminal does not sufficiently personalise the messages towards the targetted user. Target-specific and personalised phishing attacks (known as spear phishing), which have been shown to be four times more successful than a generic approach (Jagatic et al. 2007), is labour intensive. However, cost-effective spear phishing is possible using automation (Bilge et al. 2009), which researchers have demonstrated to be feasible by using machine-learning techniques to craft messages personalised to a specific user (Seymour and Tully 2016).

In the second phase of AI-supported banking fraud, AI may support the forging of an identity, including via recent advances in voice synthesis technologies (Bendel 2017). Using the classification and generation capabilities of machine learning, Adobe's software is able to learn adversarially and reproduce someone's personal and individual speech pattern from a twenty-minute recording of the replicatee's voice [p.1]. Bendel argues that AI-supported voice synthesis raises a unique threat in theft and fraud, which

> could use VoCo and Co [Adobe's voice editing and generation software] for biometric security processes and unlock doors, safes, vehicles, and so on, and enter or use them. With the voice of the customer, they [criminals] could talk to the customer's bank or other institutions to gather sensitive data or to make critical or damaging transactions. All kinds of speech-based security systems could be hacked. [p.3]

Credit card fraud is predominantly an online offence (Office for National Statistics 2016), which occurs when "the credit card is used remotely; only the credit card details are needed" (Delamaire et al. 2009). Because credit card fraud typically neither requires physical interaction nor embodiment, AI may drive fraud by providing voice synthesis or helping to gather sufficient personal details.

In the case of corporate fraud, AI used for detection may also make fraud easier to commit. Specifically,

> when the executives who are involved in financial fraud are well aware of the fraud detection techniques and software, which are usually public information and are easy to obtain, they are likely to adapt the methods in which they commit fraud and make it difficult to detect the same, especially by existing techniques (Zhou and Kapoor 2011, p. 1).

More than identifying a specific case of AIC, this use of AI highlights the risks of over-reliance on AI for detecting fraud, which may aid fraudsters. These thefts and frauds concern real-world money. A virtual world threat is whether social bots may commit crimes in massively multiplayer online game (MMOG) contexts. These online games often have complex economies, where the supply of in-game items is artificially restricted, and where intangible in-game goods can have real-world value if players are willing to pay for them; items in some cases costing in excess of US$1000 (Y. Chen et al. 2004, p. 1). So it is not surprising that, from a random sample of 613 criminal prosecutions in 2002 of online game crimes in Taiwan, virtual property thieves exploited users' compromised credentials 147 times [p.1. Fig XV] and stolen identities 52 times (Chen et al. 2005). Such crimes are analogous to the use of social bots to manage theft and fraud at large scale on social media sites, and the question is whether AI may become implicated in this virtual crime space.

Having outlined and discussed the potential for AIC as currently described in the academic literature, the time has come to analyse the solutions that are currently available. This is the task of the next section.

6.4   Conclusions

In this article, I provided a first look at the AI-Crime (AIC) as a new area of studies, which is still very much in its infancy. As a newborn area, lacking focus and direction, a question over uncertainty arises: what are the fundamentally unique and feasible threats posed by AIC? This question was answered on the basis of the classic counterfactual definition of AI and, therefore, I have focused on AI as a reservoir of autonomous smart agency that may threaten the aims and objectives of the criminal law. The threats were described area by area (in terms of specific defined crimes) and more generally (in terms of the AI qualities and issues of emergence, liability, monitoring, and psychology). The huge uncertainty over what we already know about AIC is now reduced. By reducing the unknown unknowns, it becomes clear that some AIC themes were hinted at, but never explored by the literature, and which areas of research one would expect based on criminology reports by policing organisations, but are lacking in the AIC literature. AIC, as an area of studies, is hence provided the following tentative vision for five dimensions of future AIC research, based on the surfacing of known unknowns.

Areas Understanding the areas of AIC better requires extending current knowledge, particularly concerning: the use of AI in interrogation, which was only addressed by one liability-focused paper; and theft and fraud in virtual spaces (e.g., online games with intangible assets that hold real-world value; and AAs committing emergent market manipulation, which research has seemingly only been studied in experimental simulations). The analysis revealed social engineering attacks as a plausible concern, but lacking in real-world evidence for the time being. Homicide and terrorism appear to be notably absent from the AIC literature, though they demand attention in view of AI-fuelled technologies such as pattern recognition (e.g., for identifying and manipulating potential perpetrators, or when members of vulnerable groups are unfairly targeted as suspects), weaponised drones, and self-driving vehicles; all of which may have lawful and criminal uses.

Dual-use The digital nature of AI facilitates its dual-use (Floridi 2010, p. 260; Moor 1985, p. 4), making it feasible that applications designed for legitimate uses may then be implemented to commit criminal offences. This is the case for UUVs, for example. The further AI is developed and the more its implementations become pervasive, the higher the risk of malicious or criminal uses. Left unaddressed, such risks may lead to societal rejection and excessively strict regulation of the AI-based technologies. In turn, the technological benefits to individuals and societies may be eroded as AI's use and development is increasingly constrained (Floridi and Taddeo 2016, p. 2). Such limits have already been placed on machine learning research into visual discriminators of homosexual and heterosexual men (Y. Wang and Kosinski 2017), which was considered too dangerous to release in full (i.e., with the source code and learned data structures) to the wider research community, at the expense of scientific reproducibility. Even when such costly limitations on AI releases are not necessary, as Adobe demonstrated by embedding watermarks into voice reproducing technology (Bendel 2017), external and malevolent developers may nevertheless reproduce the technology in the future. Anticipating AI's dual-use beyond the general techniques revealed in the analysis, and the efficacy of policies for restricting release of AI technologies, requires further research. This is particularly the case of the implementation of AI for cybersecurity.

Security The AIC literature reveals that, within the cybersecurity sphere, AI is taking on a malevolent and offensive role—in tandem with defensive AI systems being developed and deployed to enhance their resilience (in enduring attacks) and robustness (in averting attacks), and to counter threats as they emerge (Yang et al. 2018). The 2016 DARPA Cyber Grand Challenge was a tipping point for demonstrating the effectiveness of a combined offensive–defensive AI approach, with seven AI systems shown to be capable of identifying and patching their own vulnerabilities, while also probing and exploiting those of competing systems. More recently, IBM launched Cognitive SOC (Cognitive Security – Watson for Cyber Security | IBM 2018). This is an application of a machine-learning algorithm that uses an organisation's structured and unstructured security data, "including imprecise human language contained in blogs, articles, reports," to elaborate information about security topics and threats, with the goal of improving threat identification, mitigation, and responses. Of course, while policies will obviously play a key role in mitigating and remedying the risks of dual-uses after deployment (for example, by defining oversight mechanisms), it is at the design stage that these risks are addressed properly. Yet, contrary to a recent report on malicious AI (Brundage et al. 2018), which suggests that "one of our best hopes to defend against automated hacking is also via AI" [p.65], the AIC analysis I have conducted suggests that over-reliance on AI can be counter-productive. All of which emphasises the need for further research into AI in cybersecurity—but also into alternatives to AI.

Organisation Europol's most recent four-yearly report (Europol 2017) on the serious and organised crime threat, highlights the ways in which the type of technological crime tends to correlate with particular criminal-organisation topologies. The AIC literature indicates that AI may play a role in criminal organisations such as drug cartels, which are well-resourced and highly organised. Conversely, ad hoc criminal organisation on the dark web already takes place under what Europol refers to as crime-as-a-service. Such criminal services are sold directly between buyer and seller, potentially as a smaller element in an overall crime, which AI may fuel (e.g., by enabling profile hacking) in the future.[6] On the spectrum ranging from tightly-

---

[6] To this end a cursory search for "Artificial Intelligence" on prominent darkweb markets returned a negative result. Specifically, I checked: "Dream Market", "Silk Road 3.1", and "Wallstreet Market". The negative result is not indicative of AIC-as-a-service's absence on the darkweb, which may exist under a different guise or on more specialised markets. For example some services offer to extract personal information from a user's computer, and even if such services are genuine the underlying technology (e.g., AI-fuelled pattern recognition) remains unknown.

knit to fluid AIC organisations there exist many possibilities for criminal interaction; identifying the organisations that are essential or that seem to correlate with different types of AIC will further our understanding of how AIC is structured and operates in practice.

Developing our understanding of these dimensions is essential if we are to track and disrupt successfully the inevitable future growth of AIC. Hence, my analysis of the literature is intended to spark further research into the very serious, growing, but still relatively unexplored concerns over AIC. The sooner we understand this new crime phenomenon, the earlier we shall be able to put into place preventive, mitigating, disincentivesing, and redressing policies.

Acknowledgements This article is a shorter version of a longer one. I would like to thank the following coauthors of the longer article for their input and comments on this work: Nikita Aggarwal, Professor Luciano Floridi, and Dr. Mariarosaria Taddeo.

References

Alaieri, F., and A. Vellino. 2016. Ethical decision making in robots: Autonomy, trust and responsibility. *Lecture Notes in Computer Science (Including Subseries Lecture Notes in Artificial Intelligence and Lecture Notes in Bioinformatics) 9979 (LNAI)*: 159–168. https://doi.org/10.1007/978-3-319-47437-3_16.

Alazab, M., and R. Broadhurst. 2016. Spam and criminal activity. *Trends and Issues in Crime and Criminal Justice 526*. https://doi.org/10.1080/016396290968326.

Alvisi, L., A. Clement, A. Epasto, S. Lattanzi, and A. Panconesi. 2013. SoK: The evolution of sybil defense via social networks. *Proceedings – IEEE Symposium on Security and Privacy* (2): 382–396. https://doi.org/10.1109/SP.2013.33.

Anderson, K., and M.C. Waxman. 2013. Law and ethics for autonomous weapon systems: Why a ban won't work and how the laws of war can. *Social Science Research Network (SSRN) Electronic Journal 11*: 1–32. https://doi.org/10.2139/ssrn.2250126.

Archbold, J. 2018. *Criminal pleading, evidence and practice*. London: Sweet & Maxwell Ltd.

Ashworth, A. 2010. Should strict criminal liability be removed from all imprisonable offences? *Irish Jurist 45*: 1–21.

Bendel, O. 2017. The synthetization of human voices. *AI & SOCIETY, Online First*.

Bilge, L., T. Strufe, D. Balzarotti, E. Kirda, and S. Antipolis. 2009. All your contacts are belong to us : Automated identity theft attacks on social networks. *Www 2009*: 551–560. https://doi.org/10.1145/1526709.1526784.

Boshmaf, Y., I. Muslukhov, K. Beznosov, and M. Ripeanu. 2012. Design and analysis of a social botnet. *Computer Networks 57*: 556–578. https://doi.org/10.1016/j.comnet.2012.06.006.

Brundage, M., S. Avin, J. Clark, H. Toner, P. Eckersley, B. Garfinkel, et al. 2018. *The malicious use of artificial intelligence: Forecasting, prevention, and mitigation*. Oxford: Future of Humanity Institute.

Cath, C., S. Wachter, B. Mittelstadt, M. Taddeo, and L. Floridi. 2017. Artificial intelligence and the "good society": The US, EU, and UK approach. *Science and Engineering Ethics 24* (604): 1–23.

Chantler, A., and R. Broadhurst. 2006. Social engineering and crime prevention in cyberspace. *Queensland University of Technology 22*: 1–22.

Chen, Y., P. Chen., R. Song., and L. Korba. 2004. Online gaming crime and security issues – Cases and countermeasures from Taiwan. In *Proceedings of the 2nd annual conference on privacy, security and trust*.

Chen, Y.-C., P.S. Chen, J.-J. Hwang, L. Korba, S. Ronggong, and G. Yee. 2005. An analysis of online gaming crime characteristics. *Internet Research 15* (3): 246–261.

Chu, Z., S. Gianvecchio, H. Wang, and S. Jajodia. 2010. Who is tweeting on twitter: Human, bot, or cyborg? *Acsac 2010*: 21. https://doi.org/10.1145/1920261.1920265.

Cliff, D., and L. Northrop. 2012. The global financial markets: An ultra-large-scale systems perspective. In *Monterey workshop 2012: Large-scale complex IT systems. Development, operation and management*, 29–70. https://doi.org/10.1007/978-3-642-34059-8_2.

Cognitive Security – Watson for Cyber Security I IBM. 2018. Retrieved February 27, 2018, from https://www.ibm.com/security/cognitive.

Delamaire, L., H. Abdou, and J. Pointon. 2009. Credit card fraud and detection techniques: A review. *Banks and Bank Systems 4* (2).

Europol. 2017. *Serious and organised crime threat assessment*. European Union. Retrieved from https://www.europol.europa.eu/socta/2017/.

Ezrachi, A., and M.E. Stuck. 2016. Two artificial neural networks meet in an online hub and change the future (of Competition, Market Dynamics and Society). *Oxford legal studies research paper No. 24/2017 University of Tennessee legal studies research paper No. 323.*

Farmer, J.D., and S. Skouras. 2013. An ecological perspective on the future of computer trading. *Quantitative Finance 13* (3): 325–346. https://doi.org/10.1080/14697688.2012.757636.

Ferrara, E. 2015. *Manipulation and abuse on social media*. https://doi.org/10.1145/2749279.2749283.

Ferrara, E., O. Varol, C. Davis, F. Menczer, and A. Flammini. 2014. The rise of social bots. *Communications of the ACM 59* (7): 96–104. https://doi.org/10.1145/2818717.

Floridi, L. 2010. *The Cambridge handbook of information and computer ethics*. Cambridge: Cambridge University Press.

———. 2016. Faultless responsibility : On the nature and allocation of moral responsibility for distributed moral actions. *Royal Society's Philosophical Transactions A*: 1–22. https://doi.org/10.1098/rsta.2016.0112.

———. 2017. Digital's cleaving power and its consequences. *Philosophy & Technology 30* (2): 123–129.

Floridi, L., and J.W. Sanders. 2004. On the morality of artificial agents. *Minds and Machines 14* (3): 349–379. https://doi.org/10.1023/B:MIND.0000035461.63578.9d.

Floridi, L., and M. Taddeo. 2016. What is data ethics? *Philosophical Transactions of the Royal Society A: Mathematical, Physical and Engineering Sciences 374* (2083). https://doi.org/10.1098/rsta.2016.0360.

Floridi, Luciano, Mariarosaria Taddeo, and Matteo Turilli. 2009. Turing's imitation game: Still an impossible challenge for all machines and some judges—an evaluation of the 2008 Loebner contest. *Minds and Machines 19* (1): 145–150.

Freitas, P.M., F. Andrade, and P. Novais. 2014. Criminal liability of autonomous agents: From the unthinkable to the plausible. In *Ai approaches to the complexity of legal systems*, 145–156.

Gogarty, B., and M. Hagger. 2008. The laws of man over vehicles unmanned : The legal response to robotic revolution on sea , land and air. *Journal of Law, Information and Science 19*: 73–145. https://doi.org/10.1525/sp.2007.54.1.23.

Golder, S.A., and M.W. Macy. 2011. Diurnal and seasonal mood vary with work, sleep, and daylength across diverse cultures. *Science 333* (6051): 1878–1881. https://doi.org/10.1126/science.1202775.

Graeff, E. C. 2014. What we should do before the social bots take over: Online privacy protection and the political economy of our near future. *MIT Media Arts and Sciences*. Presented at media in transition 8: Public Media, Private Media. Place: MIT, May 5, 2013.

Grut, C. 2013. The challenge of autonomous lethal robotics to international humanitarian law. *Journal of Conflict and Security Law 18* (1): 5–23. https://doi.org/10.1093/jcsl/krt002.

Hay, G.A., and D. Kelley. 1974. An empirical survey of price fixing conspiracies. *The Journal of Law and Economics 17* (1): 13–38.

Hildebrandt, M. 2008. Ambient intelligence, criminal liability and democracy. *Criminal Law and Philosophy 2* (2): 163–180. https://doi.org/10.1007/s11572-007-9042-1.

Jagatic, T.N., N.A. Johnson, M. Jakobsson, and F. Menczer. 2007. Social phishing. *Communications of the ACM 50* (10): 94–100. https://doi.org/10.1145/1290958.1290968.
James, Gips. 1995. Towards the ethical robot. In *Android epistemology*, 243–252. Cambridge, MA: MIT Press.
Janoff-Bulman, R. 2007. Erroneous assumptions: Popular belief in the effectiveness of torture interrogation. *Peace and Conflict: Journal of Peace Psychology 13* (4): 429.
Kerr, I.R. 2004. Bots, babes and the Californication of commerce. *University of Ottawa Law & Technology Journal* 1: 287–324.
Kerr, I.R., and M. Bornfreund. 2005. Buddy bots: How turing's fast friends are under-mining consumer privacy. *Presence: Teleoperators and Virtual Environments 14*: 647–655.
Lin, T.C.W., J. Fanto, J. Fisch, J. Heminway, D. Hollis, K. Johnson, et al. 2017. The new market manipulation. *Emory Law Journal 66*: 1253.
Mackey, T.K., J. Kalyanam, T. Katsuki, and G. Lanckriet. 2017. Machine learning to detect prescription opioid abuse promotion and access via twitter. *American Journal of Public Health 107* (12): e1–e6. https://doi.org/10.2105/AJPH.2017.303994.
Marrero, Tony. 2016. Record Pacific cocaine haul brings hundreds of cases to Tampa court. *Tampa Bay Times,* September 10.
Martínez-Miranda, E., P. McBurney., and M.J. Howard. 2016. Learning unfair trading: A market manipulation analysis from the reinforcement learning perspective. *Proceedings of the 2016 IEEE Conference on Evolving and Adaptive Intelligent Systems, EAIS 2016*, 103–109. https://doi.org/10.1109/EAIS.2016.7502499.
McAllister, A. 2016. Stranger than science fiction: The rise of AI interrogation in the dawn of autonomous robots and the need for an additional protocol to the UN convention against torture. *Minnesota Law Review 101*: 2527–2573. https://doi.org/10.3366/ajicl.2011.0005.
———. 2017. Stranger than science fiction: The rise of A.I. Interrogation in the dawn of autonomous robots and the need for an additional protocol to the U.N. convention against torture. *Minnesota Law Review 101*: 2527–2573. https://doi.org/10.3366/ajicl.2011.0005.
McCarthy, J., M.L. Minsky, N. Rochester, and C.E. Shannon. 1955. *A proposal for the Dartmouth summer research project on artificial intelligence.* https://doi.org/10.1609/aimag.v27i4.1904.
Mckelvey, F., and Dubois, E. 2017. Computational propaganda in Canada: The use of political bots. *Computational Propaganda Research Project* (6): 32.
Moor, J.H. 1985. What is computer ethics? *Metaphilosophy 16* (4): 266–275.
Neff, G., and P. Nagy. 2016. Talking to bots: Symbiotic agency and the case of tay. *International Journal of Communication 10*: 4915–4931.
Nunamaker, J.F., Jr., D.C. Derrick, A.C. Elkins, J.K. Burgo, and M.W. Patto. 2011. Embodied conversational agent–based kiosk for automated interviewing. *Journal of Management Information Systems 28* (1): 17–48.
Office for National Statistics. 2016. Crime in England and Wales, year ending June 2016 – Appendix tables, (June 2017), 1–60.
Ratkiewicz, J., M. Conover., M. Meiss., B. Gonçalves., S. Patil., A. Flammini., and F. Menczer. 2011. Truthy: Mapping the spread of astroturf in microblog streams. *Proceedings of the 20th International Conference Companion on World Wide Web (WWW '11)*, 249–252. https://doi.org/10.1145/1963192.1963301
Sætenes, G.M. 2017. *Manipulation and deception with social bots : Strategies and indicators for minimizing impact,* (May).
Searle, J.R. 1983. *Intentionality: An essay in the philosophy of mind.* Cambridge: Cambridge University Press.
Seymour, J., and P. Tully. 2016. *Weaponizing data science for social engineering: Automated E2E spear phishing on Twitter.* Presented at the Black Hat USA.
Sharkey, N., M. Goodman, and N. Ross. 2010. The coming robot crime wave. *IEEE Computer Magazine 43* (8): 6–8.
Solis, G.D. 2016. *The law of armed conflict: International humanitarian law in war.* 2nd ed. Cambridge: Cambridge University Press.

Spatt, C. 2014. Security market manipulation. *Annual Review of Financial Economics* 6 (1): 405–418. https://doi.org/10.1146/annurev-financial-110613-034232.

Taddeo, M. 2017. Deterrence by norms to stop interstate cyber attacks. *Minds and Machines* (September): 10–15. https://doi.org/10.1007/s11023-017-9446-1.

Taddeo, M., and L. Floridi. 2005. Solving the symbol grounding problem: A critical review of fifteen years of research. *Journal of Experimental and Theoretical Artificial Intelligence 17* (4): 419–445.

Turing, Alan M. 1950. Computing machinery and intelligence. *Mind 59* (236): 433–460.

Twitter – Impersonation policy. 2018. Retrieved January 29, 2018, from https://help.twitter.com/en/rules-and-policies/twitter-impersonation-policy.

van de Poel, I., J.N. Fahlquist, N. Doorn, S. Zwart, and L. Royakkers. 2012. The problem of many hands: Climate change as an example. *Science and Engineering Ethics 18*: 49–67.

Van Lier, B. 2016. From high frequency trading to self-organizing moral machines. *International Journal of Technoethics (IJT) 7* (1): 34–50. https://doi.org/10.4018/IJT.2016010103.

Wang, Y., and M. Kosinski. 2017. Deep neural networks can detect sexual orientation from faces. *Journal of Personality and Social Psychology 114*: 1–47.

Wang, G., M. Mohanlal., C. Wilson., X. Wang., M. Metzger., H. Zheng., and B.Y. Zhao. 2012. *Social turing tests: Crowdsourcing sybil detection.* Retrieved from http://arxiv.org/abs/1205.3856.

Weizenbaum, J. 1976. *Computer power and human reason: From judgment to calculation.* Oxford: W. H. Freeman & Co.

Wellman, M.P., and U. Rajan. 2017. Ethical issues for autonomous trading agents. *Minds and Machines 27* (4): 609–624.

Williams, R. 2017. *Lords select committee, artificial intelligence committee, written evidence (AIC0206),* October 11. Retrieved from http://data.parliament.uk/writtenevidence/committeeevidence.svc/evidencedocument/artificial-intelligence-committee/artificial-intelligence-written/70496.html#_ftn13.

Yang, G.-Z., J. Bellingham, P.E. Dupont, P. Fischer, L. Floridi, R. Full, et al. 2018. The grand challenges of Science Robotics. *Science Robotics 3* (14): eaar7650. https://doi.org/10.1126/scirobotics.aar7650.

Zhou, W., and G. Kapoor. 2011. Detecting evolutionary financial statement fraud. *Decision Support Systems 50* (3): 570–575. https://doi.org/10.1016/j.dss.2010.08.007.

Chapter 7
The Challenges of Cyber Deterrence

Mariarosaria Taddeo

Abstract In this chapter, I analyse deterrence theory and argue that its applicability to cyberspace is limited and that these limits are not trivial. They are the consequence of fundamental differences between deterrence theory and the nature of cyber conflicts and cyberspace. The goals of this analysis are to identify the limits of deterrence theory in cyberspace, clear the ground of inadequate approaches to cyber deterrence, and define the conceptual space for a domain-specific theory of cyber deterrence, still to be developed.

7.1 Introduction

In April 2017, the foreign ministers of the G7 countries approved a 'Declaration on Responsible States Behaviour in Cyberspace' (G7 Declaration 2017). The Declaration addresses a mounting concern about international stability and the security of our societies after the fast-pace escalation of cyber attacks occurred during the past decade. In the opening statement, the G7 ministers stress their concern.

> about the risk of escalation and retaliation in cyberspace [...]. Such activities could have a destabilizing effect on international peace and security. We stress that the risk of interstate conflict as a result of ICT incidents has emerged as a pressing issue for consideration. (G7 Declaration 2017, 1)

This chapter is based on a research article published in *Philosophy & Technology*: "The Limits of Deterrence Theory in Cyberspace" (Taddeo 2017c).

M. Taddeo (✉)
Oxford Internet Institute, Digital Ethics Lab, University of Oxford, Oxford, UK

The Alan Turing Institute, London, UK
e-mail: mariarosaria.taddeo@oii.ox.ac.uk

© Springer Nature Switzerland AG 2019
C. Öhman, D. Watson (eds.), *The 2018 Yearbook of the Digital Ethics Lab*,
Digital Ethics Lab Yearbook, https://doi.org/10.1007/978-3-030-17152-0_7

Paradoxically, state actors often play a central role in the escalation of cyber attacks. State-run cyber attacks have been launched for espionage and sabotage purposes since 2003. Well-known examples include Titan Rain (2003), the Russian attack against Estonia (2006) and Georgia (2008), Red October targeting mostly Russia and Eastern European Countries (2007), Stuxnet and Operation Olympic Game against Iran (2006–2012). In 2016, a new wave of state-run (or state-sponsored) cyber attacks ranged from the Russian cyber attack against Ukraine power plant,[1] to the Chinese and Russian infiltrations of US Federal Offices,[2] to the Shamoon/Greenbag cyber-attacks on government infrastructures in Saudi Arabia.[3]

This trend will continue. The relatively low entry-cost and the high chances of success mean that states will keep developing, relying on, and deploying cyber attacks. At the same time, the AI leap of cyber capabilities—the use of AI and Machine Learning techniques for cyber offence and defence—indicates that cyber attacks will escalate in frequency, impact, and sophistication (Yang et al. 2018; Taddeo and Floridi 2018a, b).

Historically, the design and deployment of new and more effective weapons (from bombs and aircraft to chemical and nuclear weapons) have often posed the need to define new strategies to deter their use. This is also the case when considering cyber weapons. Their relatively low entry-cost and the high chances of success make cyber weapons an elective means for state and non-state actors to assert their authority, show their power, and prove their capabilities in cyberspace.

This poses serious risks of escalation, for the increasing use of cyber weapons invites frictions and tensions that may lead to the sparking of new cyber conflicts, which could intensify and jeopardise international stability and the security of our societies. For this reason, state and non-state actors, scholars, military strategists, and policy makers have increasingly stressed the need to develop cyber deterrence as a crucial step in any plan for international stability (European Union 2014; International Security Advisory Board 2014; UN Institute for Disarmament Research 2014; UK Government 2014; European Union 2015). Nonetheless, applying traditional[4] deterrence theory (henceforth simply deterrence theory) to cyberspace proves to be problematic, when not ineffective. Cyber conflicts differ radically from violent (kinetic) conflicts and define a scenario that is actually the opposite of the one for which deterrence theory was developed.

Consider Morgan's six elements of deterrence (Morgan 2003). Deterrence works in a scenario characterised by (1) a prevailing, kinetic military conflict; (2) the applicability of rational choice models to identify strategies for the involved parties; (3) positive attribution, as not problematic; (4) singular retaliation (more on this

---

[1] https://www.wired.com/2016/03/inside-cunning-unprecedented-hack-ukraines-power-grid/

[2] https://www.nytimes.com/2016/12/13/us/politics/russia-hack-election-dnc.html?_r=0

[3] https://www.symantec.com/connect/blogs/greenbug-cyberespionage-group-targeting-middle-east-possible-links-shamoon

[4] For the purposes of this chapter, I will use the expression 'traditional deterrence theory' to refer to any theory of deterrence relying on kinetic military forces, whether they be conventional or nuclear.

presently), as sufficient to inflict severe punishment to the opponent; (5) the possibility of a clear demonstration of the defender's capabilities, as well as (6) full control over retaliation. To this scenario, cyber conflicts oppose one characterised by (1) several state-run, non-kinetic cyber operations; (2) multiple (state and non-state) actors; whose (3) cost–benefit analyses vary depending on their nature; (4) non-symmetrical, multilateral interactions; (5) ever-changing dynamics; and where (6) ambiguity (rather than certainty) shapes strategies (Sterner 2011; Haggard and Simmons 1987; Jervis 1988; Libicki 2011).

The differences between kinetic and cyber scenario yield serious problems when applying deterrence theory in cyberspace. While there is a general consensus on what these problems are (for example, problems of attribution and proportionality), there is much less agreement on whether and how they can be solved (Kugler 2009; Tanji n.d.). Some suggest that these problems are unsolvable and that the nature of cyberspace is such that deterrence will ultimately be ineffective in this domain. In this vein, Lan and colleagues stress that:

> the anonymity, the global reach, the scattered nature, and the interconnectedness of information networks greatly reduce the efficacy of cyber deterrence and can even render it completely. (Lan et al. 2010, 1)

The opposite view holds that deterrence could play a crucial role in averting cyber conflicts and their escalation. The question is whether deterrence theory as it stands provides the right framework for cyber deterrence or a new theory of deterrence - "a new mind-set and changed expectations" (Sterner 2011, 62) – should be developed to address the specificity of cyber conflicts and cyberspace. I agree with this view and address this question in the rest of this chapter.

In the following sections, I will analyse the core elements of deterrence theory–attribution, defence and retaliation, and signalling–and the extent to which each of them would be effective in cyberspace. I will argue that the limits of deterrence theory in cyberspace are not trivial and indicate fundamental inconsistencies between the theory and the nature of cyber conflicts and cyberspace. The goals are to identify the limits of the application of deterrence theory to cyber conflicts (Taddeo 2016), clear the ground of inadequate approaches to cyber deterrence, and define the conceptual space for a domain-specific theory of cyber deterrence. Let me begin by focusing on the key elements of deterrence theory.

7.2 Deterrence Theory

Deterrence is a coercive strategy based on conditional threats with the goal of persuading the opponent to behave in a desirable way. It encompasses elements of control and power (both political and military), and usually has a medium- and long-term impact on the international arena. While one may trace the debate on deterrence strategies back to the 1920s and 1930s, deterrence rose to prominence only in the aftermath of World War II, when military power went from being a means to defeat the adversary, or at least of making the adversary's victory more costly than planned, to being considered as a key piece of bargaining power employed to avoid wars by means of coercion and intimidation (Possony 1946; Schelling and Affairs 1966; Schelling 1980; Brodie 1978; Zagare and Kilgour 2000; Powell 2008). It was this shift in the understanding of military power that made deterrence possible, and a particularly valuable tool in avoiding nuclear conflicts.

It follows that most of the existing analyses of deterrence have focused on East–West nuclear tensions, in particular on policies defined between the late 1940s and the 1990s to deter nuclear attacks. These analyses assumed the bipolar scenario (USA vs Soviet Union) within which deterrence seemed the obvious approach to avoid conflicts, and did not focus on "how strategic relationship of this sort might come to be established in the first place when the core [problem] was that it existed and somehow it had to be survived" (Freedman 2004, 22). Freedman's words capture the *pragmatism* of deterrence theory, which rests on three elements: (i) a context in which actors, political dynamics, interests, and military and strategic options are clearly defined; (ii) the urgency of defining effective strategies that are immediately deployable in order to avoid a nuclear conflict; both leading to (iii) deterrence theory being tantamount to deterrence policies. Indeed, the so-called 'three waves' of deterrence theory (Jervis 1988) actually identify different policy approaches endorsed by policy makers and decision makers between the late 1940s and the 1990s, rather than different theoretical stances on deterrence.

More in details, the first wave stems from Brodie's analysis of power and is based on the assumption that nuclear power was ever to be threatened and never to be deployed (Brodie 1978). The increasing reliance on rational-choice theory to maximise the bargaining of power and ensure stability characterised the second wave (Powell 2008). The third wave arose in the 1980s (Jervis 1979) and led to the dismissing of deterrence theory in international relations as a theory that was hampering, rather than encouraging, a peaceful conclusion of the Cold War. The first two 'waves of deterrence' characterise deterrence theory, and will be the focus of this section.

First and second waves deterrence strategies are modelled as follows: A believes that B is planning to attack it. In order to avoid the attack, A makes an explicit commitment to take action against B, should B decide to attack. A's commitment should be such that B is convinced that any action against A will fail, because A has the capacity either to resist or punish B, and to outweigh any prospective gains for B. B's conviction hinges on A's signalling, and credibility to act as it threatens. According to this model, we find here the three core elements of deterrence theory: the identification of the opponent (*attribution*); defence and retaliation (*deterrence strategies*); and the capability of the defender to signal credible threats (*credible signalling*) (see Fig. 6.1).

This is a minimalist model of international deterrence (D_M) defined according to deterrence theory. The D_M model is defined at a high level of abstraction (LoA, Floridi 2008) and disregards the dynamics and characteristics of specific scenarios. It assumes rational agents (a minimal assumption, given states are expected to act rationally), but it does not depend on the kinds of weapons (nuclear or conventional),

D$_M$ model

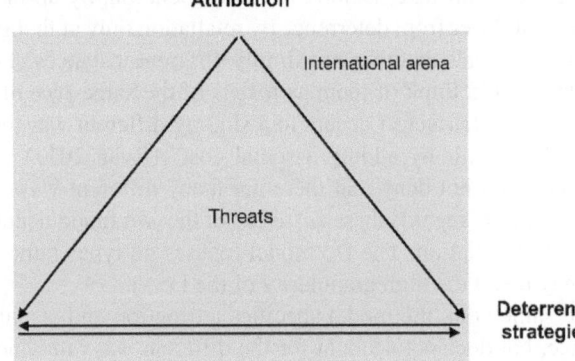

Fig. 6.1   The minimalist model (D$_M$) of international deterrence, and the dependences among its elements. (Taddeo 2017c)

the kinds of relationships between the opponents (symmetric or non-symmetric), the levels of interaction between A and B (diplomatic or not), and the scope (general or tailored) of deterrence. One may enrich the model with information about these aspects. More details would make it more complex, but would not change the dynamics and the elements identified in the D$_M$ model.

As shown in Fig. 6.1, the three core elements of the model are intertwined. Attribution is essential for deterrence, as it allows the defender to identify the target of its strategy, and also conveys a credible signal to the right opponent. At the same time, conveying a credible, coercive message to (try to) change the offender's behaviour is key in any deterrence dynamic (Libicki 2009; Bunn 2007; Jensen 2012). Indeed, effective deterrence hinges on the defender signalling its intention to use its capabilities against the offender. Credible signalling exists in a relation of mutual dependence with the deterrence strategies. That is, the chosen type of deterrence determines and underpins the content of the message and its credibility, while signalling is crucial to exploit and convey the deterring capabilities of defender, whether they be to defend or retaliate.

Of the three elements identified in the D$_M$ model, attribution and credible signalling are not controversial.[5] The identification of defence and retaliation as the two fundamental types of deterrence strategies may be more problematic, as it may be criticised for being too limited and thus undermining the completeness of the D$_M$ model. One may claim that the model should be expanded to include other deterrence strategies, which (at a first glance) do not rely on defence or retaliation, for

---

[5]Attribution may not be necessary in all instances of deterrence, for example for deterrence by defence. Some argue that when the exact source of an attack is unknown, attribution and, hence, responsibility for an attack, can be shifted to the particular state in which the attack originated (Morgan 2010; Goodman 2010). However, clear attribution remains necessary for deterrence by retaliation.

example, deterrence by association, by norms and taboos, and by entanglement. This would be a mistake. Deterrence by association, by norms and taboos, and by entanglement differ from deterrence by retaliation only at first glance. A more attentive analysis reveals that they are simply different instances of deterrence by retaliation. One should think of them as *tokens* of the same *type* of deterrence strategy, insofar as "each [strategy] occurs in a slightly different way, but all seek to punish and curb behaviour by adding a social cost" (Ryan 2017). By adopting the D_M model, one does not deny that there are many different ways to implement deterrence; one simply regards these as tokens of the two fundamental types of strategies: defence and retaliation. The D_M model focuses on types rather than on tokens (the reader may recall the high granularity of the LoA).

In the same vein, the model specifies attribution and signalling as essential for deterrence, but does not account for the different ways in which attribution can be ascertained; nor does it distinguish among the many possible modes of communication between the aggressor and the defender. Indeed, a model focusing on these aspects would be a model of specific implementations of deterrence theory, rather than a model of the theory itself.

By endorsing a high granularity LoA, the D_M model can disregard the peculiarities (while not denying their existence) of specific cases, and can focus on the necessary and sufficient elements of deterrence strategies as defined in deterrence theory. The extent to which the D_M model is valid in cyberspace will be indicative of the extent to which deterrence theory can be applied to this domain, while its limits indicate the problems that theory of cyber deterrence will have to address. Let us now consider in more detail the elements of the D_M model, starting with attribution.

7.3 Attribution and Ambiguity

Attribution is crucial both for legal and strategic reasons. Legally, attribution helps the defender to legitimise its decision to retaliate (Clark and Landau 2011; Sterner 2011). Strategically, a correct and positive attribution underpins the coercive element of deterrence, as it directs retaliation against the actual offender ensuring that the right opponent is contrasted (Iasiello 2014). If attribution is dubious or mistaken, there are chances that the defender could attack the wrong opponent, prompting new frictions and conflicts.

At the same time, uncertain attribution weakens the logic of deterrence, as it impacts the cost–benefit analysis, which underpins deterrence strategies (Libicki 2009). In particular, from the perspective of the attacker, low chances to be identified make attacks appealing and strategically advantageous, and undermine the threat of subsequent retaliation, as well as the credibility of the defender. In the eyes of the aggressor, "[the] continuing inability to attribute attacks is tantamount to an open invitation [to attack]", (Lan et al. 2010, 5). Uncertainty of attribution also heightens the risk that retaliation may be perceived either as a mistaken response or an escalation, and may hence spark new frictions and conflicts, thereby defeating the very purpose of deterrence. As Libicki stresses, in deterrence

> the lower the odds of getting caught, the higher the penalty required to convince potential attackers that what they might achieve is not worth the cost. Unfortunately, the higher the penalty [...], the greater the odds that the [retaliation] will be viewed as disproportionate— at least by third parties and perhaps even by the attacker. (Libicki 2009, 43)

These are problems faced by deterrence in cyberspace, where attribution is at best problematic, if not impossible (Libicki 2009; Goodman 2010; Jensen 2012; Haley 2013). For example, Jensen reports that most cyber attacks up until 2011 remain unattributed (Jensen 2012), and recent reports show that things are no different now.[6]

Dubious attribution is a consequence of both the distributed nature of cyberspace, which facilitates anonymity, and of the way cyber attacks are conducted. These attacks are often launched in different stages and involve globally distributed networks of machines, as well as pieces of code that combine different elements provided (or stolen) by a number of actors. This was the case, for example, with NotPetya, the malware used for an allegedly state-run cyber attack,[7] which combines a vulnerability (EternalBlue) stored by the US National Security Agency (NSA) with an ordinary remote management tool (PsExec) to access computers, gain control, and extract relevant information, such as login credentials.[8] NotPetya has inflicted serious damage worldwide and, despite recent investigations linking the attack to North Korea,[9] attribution could not be proven, precisely because the use of different tools and the particular dynamics of the attack.

In this scenario, identifying the malware, the network of infected machines, or even the country of origin of the attack is not sufficient for attribution, as it is well known that attackers can design and route their operations through third-party machines and countries with the goal of obscuring or misdirecting attribution. This has led some to maintain that uncertainty of attribution is inherent to the nature and the dynamics of cyberspace and that to solve it, we need to reengineer the internet (Kastenberg 2009; Hollis 2011). One of the proponents of this view is former director of the NSA, John Michael McConnell, who claims that "we need to reengineer the Internet to make attribution, geolocation, intelligence analysis and impact assessment—who did it, from where, why and what was the result—more manageable," (McConnell 2010). To be effective and incontestable, deterrence must be certain, severe, and immediate. Prompt, positive attribution is crucial to this end: the less positive the attribution, the more time will be necessary to respond; the less immediate is attribution, the less severe will be the defender's response. Hence, the limits of attribution in cyberspace pose serious obstacles to the deployment of

---

[6] http://www.nato.int/docu/review/2013/Cyber/timeline/EN/index.htm

[7] http://www.wired.co.uk/article/petya-malware-ransomware-attack-outbreak-june-2017

[8] https://www.theregister.co.uk/2017/06/28/petya_notpetya_ransomware/

[9] http://www.telegraph.co.uk/technology/2017/05/23/highly-likely-wannacry-cyber-attack-linked-north-korea/

effective deterrence strategies informed by deterrence theory. Recalling the D_M model, without attribution deterrence cannot function, for defence and retaliation (and indeed, signalling) are left without a target and are undermined by the inability of the defender of identify the attacker.

One could object that attribution problems are technical and therefore indicative of flaws in the design of cyberspace itself rather than shortcomings in deterrence theory per se, and that, as such, they require technical, not theoretical, solutions. According to this objection, once the appropriate technical solutions are in place, the problem of attributing cyber attacks will no longer be an obstacle to the application of deterrence theory in cyberspace.

This objection is misguided, as it fails to grasp that uncertainty of attribution is a function of the ambiguity in cyber conflicts. Ambiguity is a constitutive feature of these kinds of conflicts. It characterises attribution but also the level of violence, the assessment of their impact, and the nature of the involved actors and targets (Dipert 2013; Taddeo 2014). Thus, ambiguity should not be regarded as an unfortunate and undesired aspect of cyberspace, which nation states will want to 'correct'.

On the contrary, ambiguity is a desired feature of cyber conflicts, and technologies and practices will be developed to maintain it. This is clear when considering, for example, the reluctance of state actors to define international norms for the use of state-run cyber attacks or the failure of the U.N. Group of Governmental Experts, tasked with providing recommendations to the General Assembly on how to regulate state conduct in cyberspace. A working theory of deterrence in cyberspace has to be able to account for the ambiguity inherent to it, rather than simply trying to circumvent or ignore it.

In an offence persistent environment (more on this presently), state actors make policy decisions to protect their abilities to launch cyber attacks. 'Strategic ambiguity' is one of such decisions. According to this policy, for example, states decide neither to define nor inform the international community about their *red lines*— thresholds that once crossed would trigger state response—for non-kinetic cyber attacks (Taddeo 2012a, b), as a way to confuse the opponents about the consequences of their cyber attacks. As the US National Intelligence Officer for Cyber Issues officer put it:

> Currently most countries, including ours, don't want to be incredibly specific about the red lines for two reasons: You don't want to invite people to do anything they want below that red line thinking they'll be able to do it with impunity, and secondly, you don't want to back yourself into a strategic corner where you have to respond if they do something above that red line or else lose credibility in a geopolitical sense.

However, by fostering ambiguity, state actors also leave open for themselves a wider room for manoeuvring. Strategic ambiguity allows state actors to deploy cyber attacks for military, espionage, sabotage, and surveillance purposes without being constrained by their own policies or international red lines. This makes ambiguity a dangerous choice, one that is strategically risky and politically misleading.

The risks come with the cascade effect following the absence of clear thresholds for cyber attacks. The lack of thresholds facilitates a proliferation of offensive strategies. This in turn favours an international cyber arms race and the weaponisation of cyberspace, which ultimately spurs the escalation of cyber attacks. In parallel, while seeking to maintain uncertainty about red lines to deter prospective cyber attacks, strategic ambiguity actually ends up leaving unbounded (state and non-state run) non-kinetic cyber attacks, which are indeed the great majority of cyber attacks.

As we shall see in the rest of this chapter, when applied in cyberspace, all three core elements of deterrence theory identified in the D_M model face serious problems, indicating that it would be problematic to design deterrence strategies in cyberspace that rely on this theory even if attribution was not a problem and state actors abandoned strategic ambiguity. Deterrence theory simply does not account properly for the dynamics of cyber conflicts, the nature of cyberspace, and the malleability of cyber weapons (Taddeo 2017a, b). The next section will focus on each of these aspects.

7.4 Deterrence Strategies: Defence and Retaliation

Both deterrence by defence and by retaliation include elements of coercion and control, although to different extents. Deterrence by defence is essentially concerned with controlling the impact of an attack either by preventing (dismounting) it or by rendering it ineffective (ensuring that, even if it breaches the defences, the attack does not reach its intended target). Both aspects act as deterrent as they ensure that new attacks will inevitably fail. Effective defence has also a coercive element, insofar as by discouraging or thwarting an otherwise successful attack, the defendant forces the opponent to change its behaviour.

Deterrence by retaliation is essentially coercive. It rests on the (threat of) use of force to change the offensive plan of the opponent. State A launches, or threatens to launch, a counter strike that imposes a cost on State B and outweighs B's benefit from the initial attack. In view of these likely costs, State B decides not to attack. Retaliation also has an element of control, as full control of the impact and the scope of the retaliation is crucial to avoid breaches of proportionality and risks of escalation.

Deterrence strategies based on either defence or retaliation (or a combination of both) are problematic, if not ineffective, when deployed in cyberspace.

7.4.1 Defence in Cyberspace

Anticipating defence is guaranteed to be ineffective as a deterrence strategy in cyberspace, because cyber defence mechanisms have little control over cyber attacks. This deprives defence of any strategic power (Taddeo 2017a, b), and transforms it into a means of ensuring resilience of information systems, rather than a means to deter new attacks. Let me unpack this analysis.

Defence in cyberspace is porous in nature (Morgan 2010); every system has its security vulnerabilities, and identifying and exploiting them is simply a matter of time, means, and determination. This makes ephemeral even the most sophisticated defence mechanisms, thus limiting their potential to deter new attacks (Taddeo 2017a, b, c). At the same time, even when successful, cyber defence does not lead to a strategic advantage, insofar as dismounting a cyber attack very rarely leads to the ultimate defeating of an adversary. This creates an environment of *persistent offence* (Harknett and Goldman 2016), where attacking is tactically and strategically more advantageous than defending.

An 'offence persistent' environment differs from an 'offence dominant' environment, in that defence is under constant stress, but is not superfluous, and the success of the offence is not a given. In an offence persistent environment, defence can achieve tactical and operational success in the short-term if it can adjust constantly to the means of attack, but it cannot win strategically. Offence will persist and the interactions with the enemy will remain constant (Harknett and Goldman 2016). In this kind of environment, deterrence by defence is guaranteed to be ineffective, as defence does not discourage attackers from their intention to offend. This is even more so in cyberspace, where uncertainty of attribution, low entry cost of attacks, and the inherently vulnerable nature of information systems encourage attackers to test defences.

In cyberspace, defence remains salient and necessary, but primarily as a means to guarantee the resilience of a system once an attack has been launched (and also after it has breached the system), rather than as a means of deterring attackers (Bologna et al. 2013; Bendiek and Metzger 2015). Cyber defence, then, is more akin to safety engineering, in that it mitigates and manages the risk following an attack (Libicki 1997; Rattray 2009), rather than avoiding them.

7.4.2 Retaliation in Cyberspace

Given the guaranteed ineffectiveness of defence as a deterrence strategy, and offence persistency in cyberspace, state actors focus their attention on developing cyber deterrence by retaliation. As Croston stresses, "the goal for major powers should not be the futile hope of developing a perfect defensive system of cyber deterrence, but rather the ability to instil deterrence based on a mutually shared fear of an offensive threat," (Crosston 2011). Approaches to deterrence by retaliation in cyberspace often make reference to nuclear deterrence models. Some consider mutual assured destruction (MAD) a viable strategy to shape cyber deterrence, given its potential to limit the freedom of major political actors to attack each other: "By capitalizing on this shared vulnerability to attack and propagandizing the open build-up of offensive capabilities, there would arguably be a greater system of cyber deterrence keeping the virtual commons safe," (Crosston 2011). This approach rests on the idea that analyses and practices of nuclear deterrence can shed light on cyber deterrence (Owens et al. 2009). Nye outlines this idea quite clearly: "There are some important nuclear-cyber strategic rhymes, such as the superiority of offense over defense, the potential use of weapons for both tactical and strategic purposes, the possibility of first- and second-use scenarios, the possibility of creating automated responses when time is short, the likelihood of unintended consequences and cascading effects," (Nye 2011, 22–23). Aside from the guaranteed ineffectiveness of defence as a deterrent, which is indeed an aspect peculiar to both nuclear and cyber conflicts, the rest of the similarities listed by Nye are too generic to characterise nuclear and cyber conflicts as equivalent. They can, actually, be used to describe many types of modern warfare; air and marine warfare, for example, meet all the requirements he lists.

An attentive analysis unveils that nuclear and cyber conflicts differ radically in several crucial aspects. Differences range from clarity of attribution, the destructive power of the attacks, the military capabilities of the opponents, as well as the nature of the involved actors (Sterner 2011; Taddeo 2016). These differences shape diverging deterrence strategies. Nuclear deterrence is singular and symmetric (Libicki 2009), while cyber deterrence is repeatable and non-symmetric. Nuclear deterrence is singular, as by the time a nuclear attack and retaliation have run their course both parties are likely destroyed and there is no chance for the offender to counter retaliate. At the same time, nuclear deterrence works only among actors enjoying symmetric military power: a state with no nuclear capacity could not deter a nuclear power on those terms.

Unlike nuclear deterrence, cyber deterrence is repeatable, as non-kinetic retaliations are unlikely to defeat the opponent definitively, let alone pose ultimate threats (Libicki 2009; Taddeo 2017a, b), thus leaving the aggressor able to counter retaliate and favouring multiple interactions between defender and offender. Early analyses (Libicki 2009) maintain that cyber deterrence between states is symmetric, as it occurs among peers and the defender and offender are assumed to share the same strategic ground. This is only partially correct, as this view overlooks more complex scenarios, where the defender may have inferior cyber capabilities and may use (proportionate) kinetic means to retaliate, or where the offender relies on cyber means to attack an opponent with superior kinetic means. The non-symmetric use of cyber capabilities has been acknowledged in a leaked NSA report (NSA 2013), which recognises that "[c]yberattacks offer a means for potential adversaries to overcome overwhelming U.S. advantages in conventional military power and to do so in ways that are instantaneously and exceedingly hard to trace," (NSA 2013, 3). Even when focusing only on state actors, it is not possible to assume symmetry between the cyber capabilities of the defender and offender. For this reason, I argue that cyber deterrence is non-symmetric, as the defender and the offender may or may not enjoy the same kinetic and cyber capabilities. This is crucial, as it means that in deciding whether or not to retaliate, the defender will have to consider the possibility of both kinetic and non-kinetic counter retaliation and, hence, of escalation. The two couples 'singular and symmetric' and 'repeatable and non-symmetric' indicate that nuclear and cyber deterrence are not related and, thus, that analogies between nuclear and cyber deterrence are unwarranted.

Deterrence strategies are also heavily determined by the nature of the threats that they pose. Nuclear conflicts pose existential threats—a nuclear attack is likely to destroy both opponents, their infrastructures, and populations; while cyber conflicts do not. Indeed, cyber conflicts proliferate because they pose *non-existential* threats (Taddeo 2012a, b). This difference is critical when considering deterrence. In nuclear deterrence, the existential nature of the threats justifies and makes credible MAD strategies. In contrast, cyber deterrence affords to the defender a whole range of possible strategies —from in-kind retaliation and economic sanctions, to diplomatic measures and proportionate kinetic responses—because of the non-existential nature of (non-kinetic) cyber threats. These options are lost when modelling cyber deterrence in analogy to nuclear deterrence.

7.4.3   Control and Risks of Cyber Deterrence by Retaliation

Even when not informed by analogies with nuclear strategies, retaliation as identified in deterrence theory raises serious problems when applied in cyberspace. Unlike with defence, deterrence by retaliation is not guaranteed to be ineffective in this domain. Indeed, in an offence persistent environment like cyberspace, retaliation can actually be a successful strategy. However, when deployed in cyberspace, the nature of cyber weapons and of cyber conflicts undermines the control element of retaliation, making it a hazardous choice for deterrence.

Retaliation is coupled with the risk of escalation. This risk is amplified when retaliation occurs in a non-symmetric scenario, where the opponent may lack cyber capabilities and instead counter-retaliates using kinetic means. Control over the weapons used and the impact of the resulting retaliation is crucial to avoid escalation. In cyberspace, however, this control is limited given the *malleability* of cyber weapons.

Cyber weapons are malleable insofar as they can be accessed, stored, combined, repurposed, and re-deployed much more easily than was ever possible with other kinds of military capability (Schneier 2017). Repurposing or re-deploying state-designed or state-owned malware is not too rare an event. It happened in 2011 with Stuxnet. Despite being designed to target specific configuration requirements of Siemens software installed on Iranian nuclear centrifuges, the worm was eventually released on the Internet and infected systems in Azerbaijan, Indonesia, India, Pakistan, and the US.[10] Even more worryingly, the vulnerability that Stuxnet exploited has been used for at least 6 years to weaponise Angler, one of the most infectious malwares used by cyber criminals to target online banking websites.[11] In

---

[10] https://www.symantec.com/security_response/writeup.jsp?docid=2010-071400-3123-99

[11] https://www.theregister.co.uk/2016/05/09/sixyearold_patched_stuxnet_hole_still_the_webs_biggest_killer/

the same vein, in 2017 two major cyber attacks, WannaCry and NotPetya, repurposed an exploit (EternalBlue) stolen from the NSA.[12]

The chances of a cyber weapon causing more damage than originally planned increase when considering the ever-more likely deployment of 'counter autonomy' systems for national defence. These are machine-learning systems that are able to identify and target autonomously vulnerabilities in other systems, while also isolating and patching their own.[13] As machine-learning systems learn and evolve in an unsupervised way, their use for defence purposes poses concrete risks of unforeseen, disproportionate damage (Cath et al. 2017).

The malleability of cyber weapons erodes the control element of retaliation in cyberspace and, in so doing, makes retaliation a dangerous strategic choice, with the potential for disastrous cascade effects. Weak control over the impact of retaliation could lead to a breaching of proportionality, in turn triggering self-defence and prompting escalation. Ensuring control over retaliation is essential to avoid these unintended effects, and respecting proportionality is crucial to this end.

There is a general consensus that the principle of proportionality applies in the case of cyber deterrence and that it does not require an in-kind response (Libicki 2009; Jensen 2009; Goodman 2010; Iasiello 2014). Hence, a consensus that retaliation to a cyber attack could use cyber or kinetic means (or a combination of them), as long as the response is comparable to the impact of the initial attack and does not equate to an escalation (Hathaway and Crootof 2012).

However, determining the impact of a non-kinetic cyber attack is problematic. Proportionality prescribes that retaliation should equate to the actual (and not just the discovered) damage suffered by the defender. This can be a serious hurdle in cyberspace, where "very little [...] can be inferred about unseen activities (which cannot be measured) from those that are seen (which can be measured)," (Libicki 2009, 103). At the same time, even when the attack is detected and its impact is clear, it can be difficult to assess the value and type of damage, and therefore an appropriate response. As Harknett and Goldman note:

> If an attack reduces no buildings to rubble and kills no one directly, but destroys information, what is the response? We tend to think about information as intangible, but the loss of information can have tangible personal, institutional, and societal costs. What credibly can be placed at risk that would dissuade a state from contemplating such an attack? (Harknett and Goldman 2016)

The questions that they pose hinge on what I have described in a previous article (Taddeo 2014) as an *ontological gap* between Just War Theory and cyber conflicts. This gap refers to the difference between the ontology assumed by Just War Theory—which is centred on human beings, tangible objects, and kinetic conflicts causing physical damage and bloodshed—and the nature of (artificial) agents, (digital) targets, and (non-kinetic) cyber conflicts (Taddeo 2012a, b). Because of this

---

[12] https://www.forbes.com/sites/thomasbrewster/2017/05/12/nsa-exploit-used-by-wannacry-ransomware-in-global-explosion/#3f04a279e599

[13] https://fas.org/irp/agency/dod/dsb/autonomy-ss.pdf;   https://www.darpa.mil/program/cyber-grand-challenge

gap, it is problematic to apply Just War Theory principles to cyber conflicts. In the case of proportionality and deterrence, in particular, we face the danger that uncertainties in assessing proportionality may justify self-defence and facilitate escalation.

While proportionality still remains a valid and desirable principle to regulate cyber conflicts, its application poses serious problems that can only be resolved once the ontological gap is overcome. In turn, addressing the gap will require an ethical framework for the regulation of cyber conflicts that takes account of the moral stance of informational entities, like artificial agents and digital targets, as well as that of human beings and tangible objects (Taddeo 2012a, b; Floridi 2013; Floridi and Taddeo 2014). Without such a framework, attempts to define deterrence strategies in cyberspace that are able to respect and make sense of proportionality are doomed to failure.

In kinetic scenarios, defence and retaliation strategies offer the perfect balance between control of response and coercion that ultimately allows the defender to show its power and deter the offender. In cyberspace, this balance is not achievable, as both defence and retaliation lack control. Recalling the D_M model, neither of these two types of deterrence strategies works in cyberspace. However, there is an important difference to be noted: while defence strategies are guaranteed to be ineffective in an offence persistent environment, like cyberspace, retaliation could be a viable strategic choice (within the limits posed by attribution).

Nonetheless, to be successful, retaliation needs to be reconsidered to ensure that, while remaining essentially about coercion, it can rely on strong control mechanisms that ensure a proportionate response. This requires addressing the ontological gap and overcoming the limits of Just War Theory in cyberspace. The alternative is to model retaliation in cyberspace using MAD strategies, but this is more likely to lead to escalation than it is to deter new conflicts.

7.5   Credible Signalling

A defender deters prospective attackers by signalling to the attacker its awareness of the offender's plans and the envisaged response, should the plan be implemented. Without this signalling, deterrence would not be possible. Iasiello, for example, notes that retaliation becomes ineffective and can be misinterpreted if the defender is not able to convey a credible signal of its intentions (Iasiello 2014).

As shown in the D_M model, signalling is only effective insofar as it conveys a coercive message (threat) and, thus, it depends on the deployment of an appropriate deterrence strategy (see Fig. 6.1). The message has to be credible. The credibility of the message hinges on the reputation of the defender to follow through on its threats (Freedman 2004). Indeed, reputation is a central aspect of deterrence theory. Famously, Schelling stressed that "'Face' is one of few things worth fighting over [...] 'face' is merely the interdependence of a country's commitments; it is a country's reputation for actions, the expectations other countries have about its behaviour", (Freedman 2004, 53). In kinetic scenarios, reputation is gained by showcasing a state's military capabilities—military parades, and deployment of soldiers or ships on the borders of the offending state typically serve this purpose—as well as by showing ability to resolve (to deter or defeat the opponent) over time. To some extent, the same also holds true in cyberspace, where a state's reputation also refers to a state's past interactions in this domain, its known cyber capabilities to defend and offend, as well as its overall reputation in resolving conflicts. One caveat is that a state's reputation in cyberspace may not necessary correspond to its actual capabilities in this domain, as states are reluctant to circulate information about the attacks that they receive. In the medium- and long-term, this may make signalling less credible, and thus more problematic, than in other domains of warfare.

Signalling can be either general or tailored. General signalling conveys a message about the overall deterrence strategy to the rest of the international arena, through open statements released by a state conveying information about its approaches, commitments, and capabilities. Although it may be problematic in some circumstances, general signalling in cyberspace is not impossible. The reference to the ability to resort to the "full range of tools available to the United States" in the US cyber strategy document (US Government 2015, 14), as well as mention of the Active Cyber Defence capabilities in the UK equivalent (UK Government 2015) serve precisely this purpose. In both cases, general signalling is credible as it rests on the reputation that the US and UK have in cyberspace, as well as in the international arena.

Tailored signalling—the conveying of a threat to a specific offender indicating the possible targets of retaliation—is more problematic than general signalling, and constitutes a significant obstacle to delivering effective deterrence strategies in cyberspace. This kind of signalling is effective if attribution is certain. If the defender has not identified the offender correctly, tailored signalling can be counterproductive given it may be directed to the wrong actor. Tailored signalling also requires a careful fine-tuning in order not to expose the defender's capabilities and assets, especially when the defender is considering retaliation in-kind.

The risks are multiple and range from exposing knowledge about the opponent's cyber assets, which would imply that the defender has also run cyber operations (sabotage or espionage) against the opponent, to revealing the defender's assets and strategies, which may expose and therefore render futile its cyber capabilities, such as zero-day exploits (for example). At the same time, too vague a signalling would undermine the credibility of the threats and the success of deterrence. Two alternatives follow for cyber deterrence: signalling could become increasingly decoupled from reputation, though thereby also weakening the coercive nature of deterrence; or deterrence could become less about signalling as a way of alerting the opponent, and more about *demonstrating* the capabilities and intentions of the defender. But this will pose increasing risks for escalation. Both scenarios undermine the chances of successful deterrence strategies.

7.6   Conclusion

The limits of deterrence theory in cyberspace are not trivial, and they follow the fundamental differences between kinetic and cyber conflicts outlined in this chapter. These differences cannot be disregarded when defining deterrence strategies in cyberspace. As have I argued elsewhere (Taddeo 2017b), understanding these differences—and identifying their impact on international relations and on military strategies—is a preliminary and necessary step to any attempt to develop strategies for cyber deterrence. For this reason, we must recognise the limits of approaching cyber deterrence by analogy with kinetic conflicts, and move past them. As Betz and Stevens put it: "It is little wonder that we attempt to classify [...] the unfamiliar present and unknowable future in terms of a more familiar past, but we should remain mindful of the limitations of analogical reasoning in cyber security," (Betz and Stevens 2013). Analogies can be powerful, for they inform the way in which we think and constrain ideas and reasoning within a conceptual space (Wittgenstein 2009). However, if the conceptual space is not the right one, analogies become misleading and detrimental for any attempt to develop innovative and in-depth understanding of new phenomena, and they should be abandoned altogether. When the conceptual space is the right one, analogies are at best a step on Wittgenstein's ladder and need to be left behind once they have taken us to the next level of the analysis. In the best scenario, this is the case of the analogies between traditional and cyber deterrence.

Once we have abandoned any unhelpful analogies with traditional deterrence, a theory of cyber deterrence will have to develop an original approach to address the specificities of cyber conflicts. These span three areas: conceptual, normative, and regulative.

Conceptually, the non-kinetic nature of cyber conflicts has redefined our understanding of key notions such as harm, violence, target, combatants, weapons, attack, and political power, and a theory of cyber deterrence will be successful only insofar as it rests on a clear grasp of these concepts. At the same time, the ontological gap and the limits of Just War Theory, identified in Sect. 4.3, highlight the absence of clear ethical guidance in shaping cyber deterrence. There is a pressing need to define ethical principles to ensure the deployment of deterrence strategies able to respect individual rights (Taddeo 2013; Taddeo and Glorioso 2016), avoid unnecessary violence and bloodshed, and ultimately foster a peaceful geopolitical environment.

Deterrence must be deployed in accordance with international humanitarian laws, and conceptual and normative problems pose sever hindrances to the regulation of cyber deterrence. However, as these laws were defined with respect to kinetic conflicts, it is unclear to what extent, if at all, they offer the right guidance to regulate cyber deterrence (Taddeo 2016), and therefore to guarantee the stability of international relations of current and future information societies.

References

Bendiek, A., and T. Metzger. 2015. *Deterrence theory in the cyber-century: Lessons from a state-of-the-art literature review*, Lecture notes in informatics (LNI), 553–570. Bonn: Gesellschaft Fur Informatik.

Betz, D.J., and T. Stevens. 2013. Analogical reasoning and cyber security. *Security Dialogue* 44 (2): 147–164. https://doi.org/10.1177/0967010613478323.

Bologna, S., A. Fasani, and M. Martellini. 2013. From fortress to resilience. In *Cyber security: Deterrence and IT protection for critical infrastructures*, ed. M. Martellini, 53–56. Heidelberg: Springer.

Brodie, B. 1978. The development of nuclear strategy. *International Security* 2 (4): 65–68.

Bunn, M.E. 2007. *Can deterrence be tailored?* Strategic Forum, Number 225, January 2007. Washington, DC: Institute for National Strategic Studies, National Defense University.

Cath, C., S. Wachter., M. Taddeo., and L. Floridi. 2017. Artificial intelligence and the "good society": The US, EU, and UK approach. *Science and Engineering Ethics*, March. https://doi.org/10.1007/s11948-017-9901-7.

Clark, D., and S. Landau. 2011. Untangling attribution. *Harvard National Security Journal* 2011 (2): 25–40.

Crosston, M. 2011. World gone cyber MAD: How "mutually assured debilitation" is the best hope for cyber deterrence. *Strategic Studies Quarterly* 50 (1): 100–116.

Dipert, R. 2013. The essential features of an ontology for cyberwarfare. In *Conflict and cooperation in cyberspace*, ed. Panayotis Yannakogeorgos and Adam Lowther, 35–48. Taylor & Francis. http://www.crcnetbase.com/doi/abs/10.1201/b15253-7.

European Union. 2014. *Cyber defence in the EU: Preparing for cyber warfare? – think tank.* Brussels. http://www.europarl.europa.eu/thinktank/en/document.html?reference=EPRS_BRI(2014)542143.

———. 2015. *Cyber diplomacy: EU dialogue with third countries – think tank.* Brussels. http://www.europarl.europa.eu/thinktank/en/document.html?reference=EPRS_BRI(2015)564374.

Floridi, L. 2008. The method of levels of abstraction. *Minds and Machines* 18 (3): 303–329. https://doi.org/10.1007/s11023-008-9113-7.

———. 2013. *The ethics of information.* Oxford: Oxford University Press.

Floridi, L., and M. Taddeo, eds. 2014. *The ethics of information warfare*, Law, governance and technology series. Vol. 14. Heidelberg: Springer.

Freedman, L. 2004. *Deterrence.* Cambridge, UK/Malden: Polity Press.

G7 Declaration. 2017. *G7 declaration on responsible state behavior in cyberspace.* Lucca. http://www.mofa.go.jp/files/000246367.pdf.

Goodman, W. 2010. Will Goodman, Cyber deterrence: Tougher in theory than in practice?, *Strategic Studies Quarterly* Fall: 102–35.

Haggard, S., and B.A. Simmons. 1987. Theories of international regimes. *International Organization* 41 (03): 491.

Haley, C. 2013. A theory of cyber deterrence. *Georgetown Journal of International Affairs*, February. http://journal.georgetown.edu/a-theory-of-cyber-deterrence-christopher-haley/.

Harknett, R.J., and E.O. Goldman. 2016. The search for cyber fundamental. *Journal of Information Warfare* 15 (2): 81–88.

Hathaway, O., and R. Crootof. 2012. The law of cyber-attack. *California Law Review* 100 (1–2012): 817–886.

Hollis, D.B. 2011. An E-SOS for cyberspace. *Harvard International Law Journal.* 52 (373): 374–375.

Iasiello, E. 2014. Is cyber deterrence an illusory course of action? *Journal of Strategic Security* 7 (1): 54–67.

International Security Advisory Board. 2014. *A framework for international cyber stability.* United States Department of State. http://goo.gl/azdM0B.

Jensen, E.T. 2009. Cyber warfare and precautions against the effects of attacks. *Texas Law Review* 88 (1533): 1534–1569.

———. 2012. *Cyber deterrence*. SSRN scholarly paper ID 2070438. Rochester: Social Science Research Network. https://papers.ssrn.com/abstract=2070438.

Jervis, R. 1979. Deterrence theory revisited. *World Politics* 31 (2): 289–324. https://doi.org/10.2307/2009945.

———. 1988. Realism, game theory, and cooperation. *World Politics* 40 (3): 317–349. https://doi.org/10.2307/2010216.

Kastenberg, J.E. 2009. Changing the paradigm of internet access from government information systems: A solution to the need for the DoD to take time-sensitive action on the Niprnet. *Air Force Law Review* 64: 175.

Kugler, R. 2009. Deterrence of cyber attacks. In *Cyberpower and national security*, ed. Franklin Kramer, Stuart Starr, and Larry Wentz, 309–342. Washington, DC: National Defense University.

Lan, Tang, Zhang Xin, Dmitry Grigoriev Harry Raduege Jr., Pavan Duggal, and Stein Schjølberg. 2010. *Global cyber deterrence views from China, the U.S., Russia, India, and Norway*. EastWest Institute.

Libicki, M.C. 1997. *Defending cyberspace and other metaphors*. Washington, DC: National Defense Univ/National Strategic Studies.

Libicki, Martin. 2011. The strategic uses of ambiguity in cyberspace. *Military and Strategic Affairs* 3 (3): 3–10.

———. 2009. *Cyberdeterrence and cyberwar*. Product Page. http://www.rand.org/pubs/monographs/MG877.html.

McConnell, M. 2010. *Mike McConnell on how to win the cyber-war we're losing*, 28 February 2010. http://www.washingtonpost.com/wp-dyn/content/article/2010/02/25/AR2010022502493.html.

Morgan, P.M. 2003. *Deterrence now*, Cambridge studies in international relations 89. Cambridge, UK/New York: Cambridge University Press.

———. 2010. Applicability of traditional deterrence concepts and theory to the cyber realm. In *Proceedings of a workshop on deterring cyberattacks: Informing strategies and developing options for U.S. policy*, 55–76. Washington, DC: National Academic Press.

NSA. 2013. A strategy for surveillance powers. *The New York Times*. http://www.nytimes.com/interactive/2013/11/23/us/politics/23nsa sigint strategy document.html.

Nye, J.S. 2011. Nuclear lessons for cyber security? *Strategic Studies Quarterly* 5 (4): 11–38.

Owens, W.A., Kenneth W. Dam., H. Lin., National Research Council (U.S.), National Research Council (U.S.), National Research Council (U.S.), eds. 2009. *Technology, policy, law, and ethics regarding U.S. Acquisition and use of cyberattack capabilities*. Washington, DC: National Academies Press.

Possony, S.T. 1946. Atomic power and world order. *The Review of Politics* 8 (4): 533–535.

Powell, R. 2008. *Nuclear deterrence theory: The search for credibility*, Digitally printed version. Paperback Re-Issue. Cambridge: Cambridge University Press.

Rattray, G.J. 2009. An environmental approach to understanding cyberpower, in Kramer, Cited, 253–274, Esp. 256. In *Cyberpower and National Security*, ed. Stuart S. Kramer and Lerry K. Wentz, 253–74. Washington, DC: National Defense UP.

Ryan, N.J. 2017. Five kinds of cyber deterrence. *Philosophy & Technology*, January. https://doi.org/10.1007/s13347-016-0251-1.

Schelling, T.C. 1980. *The strategy of conflict: [With a new preface]*. Cambridge, MA: Harvard University Press.

Schelling, T.C., and Harvard University Center for International Affairs. 1966. *Arms and influence*. New Haven: Yale University Press.

Schneier, B. 2017. Why the NSA makes us more vulnerable to cyberattacks. *Foreign Affairs*, 30 May 2017. https://www.foreignaffairs.com/articles/2017-05-30/why-nsa-makes-us-more-vulnerable-cyberattacks.

Sterner, E. 2011. Retaliatory deterrence in cyberspace. *Strategic Studies Quaterly* 5 (1): 65–80.

Taddeo, M. 2012a. Information warfare: A philosophical perspective. *Philosophy & Technology* 25 (1): 105–120. https://doi.org/10.1007/s13347-011-0040-9.

———. 2012b. 'An analysis for a just cyber warfare'. In 2012 4th International Conference on Cyber Conflict (CYCON 2012), 1–10.

———. 2013. Cyber security and individual rights, striking the right balance. *Philosophy & Technology* 26 (4): 353–356. https://doi.org/10.1007/s13347-013-0140-9.

———. 2014. Information warfare: The ontological and regulatory gap. *Newsletter on Philosophy and Computers* 14 (1 (fall 2014)): 13–20.

———. 2016. On the risks of relying on analogies to understand cyber conflicts. *Minds and Machines* 26 (4): 317–321. https://doi.org/10.1007/s11023-016-9408-z.

———. 2017a. Cyber conflicts and political power in information societies. *Minds and Machines* 27 (2): 265–268. https://doi.org/10.1007/s11023-017-9436-3.

———. 2017b. Deterrence by norms to stop interstate cyber attacks. *Minds and Machines*, September. https://doi.org/10.1007/s11023-017-9446-1.

———. 2017c. The limits of deterrence theory in cyberspace. *Philosophy & Technology*, October. https://doi.org/10.1007/s13347-017-0290-2.

Taddeo, M., and L. Glorioso. 2016. Regulating cyber conflicts and shaping information societies. In *Ethics and policies for cyber operations, philosophical studies series*. Berlin/Heidelberg: Springer.

Taddeo, Mariarosaria, and Luciano Floridi. 2018a. How AI can be a force for good. *Science* 361 (6404): 751–752.

———. 2018b. Regulate artificial intelligence to avert cyber arms race. *Nature* 556 (7701): 296–298.

Tanji, M. n.d. *Deterring a cyber attack? Dream on... WIRED.* Accessed 15 July 2017. https://www.wired.com/2009/02/deterring-a-cyb/.

UK Government. 2014. *Deterrence in the twenty-first century: Government response to the committee's eleventh report.* http://www.publications.parliament.uk/pa/cm201415/cmselect/cmdfence/525/52504.htm.

———. 2015. *National Security Strategy 2016–2021.* London: HM Government. https://www.gov.uk/government/uploads/system/uploads/attachment_data/file/567242/national_cyber_security_strategy_2016.pdf.

UN Institute for Disarmament Research. 2014. *Cyber stability seminar 2014: Preventing cyber conflict.* Geneva: UN Institute for Disarmament Research.

US Government. 2015. The Department of Defense cyber strategy. Washington, DC, USA.

Wittgenstein, L. 2009. *Philosophical investigations*, eds. P.M.S. Hacker and Joachim Schulte, Rev. 4th ed. Chichester/Malden: Wiley-Blackwell.

Yang, Guang-Zhong, Jim Bellingham, Pierre E. Dupont, Peer Fischer, Luciano Floridi, Robert Full, Neil Jacobstein, et al. 2018. The grand challenges of science robotics. *Science robotics* 3 (14): eaar7650. https://doi.org/10.1126/scirobotics.aar7650.

Zagare, F.C., and D.M. Kilgour. 2000. *Perfect deterrence*, Cambridge studies in international relations 72. Cambridge, UK/New York: Cambridge University Press.

Chapter 8
Internet Governance and Human Rights: A Literature Review

Corinne Cath

Abstract This article presents an integrative literature review on internet governance and human rights. In it, I consider two arguments. First, I build on existing literature that internet governance is political to suggest it should consider including a broader set of internet design and governance questions beyond the current research focus taken. Second, that research on internet governance and human rights is comprehensive but skewed towards covering particular technologies, rights, actors, and means for enabling human rights online. This is, in part, the result of how researchers define the field and who they consider part of their epistemic community. Concluding, I present various suggestions for further research and policy making.

8.1 Introduction

On September 21, 2016, Republican Presidential Candidate Donald Trump issued a press statement opposing 'President Obama's plan to surrender American internet control to foreign powers' (Trump Campaign, 2016).[1] This press statement came in response to the political discussion about the US government's decision to relinquish its control of the stewardship role of the Internet Assigned Names and Numbers Authority (IANA). This authority is part of the technical functioning of the internet as managed by the internet Corporation for Assigned Names and Numbers (ICANN), which maintains the Domain Name System (DNS). The DNS matches domain names with Internet Protocol (IP) addresses, ensuring one can find

---

[1] Trump Campaign. (2016). Trump opposes president Obama's plan to surrender American internet control to foreign powers. Retrieved from https://web.archive.org/web/20170430195910/https://www.donaldjtrump.com/press-releases/donald-j.-trump-opposes-president-obama-plan-to-surrender-american-internet

C. Cath (✉)
Oxford Internet Institute, Digital Ethics Lab, University of Oxford, Oxford, UK

The Alan Turing Institute, London, UK
e-mail: ccath@turing.ac.uk

© Springer Nature Switzerland AG 2019
C. Öhman, D. Watson (eds.), *The 2018 Yearbook of the Digital Ethics Lab*,
Digital Ethics Lab Yearbook, https://doi.org/10.1007/978-3-030-17152-0_8

content online. Contrary to popular belief, the US was not giving up 'American control' over the internet. Nor would it be handing the internet over to 'repressive states'. Because no single state ever controlled the internet (Naughton 1999; Mueller 2010). Rather, the transition meant that the unilateral American oversight over IANA transferred to ICANN's multistakeholder community.

The IANA transition was not the first time that the technical governance of the internet became a proxy for political struggle. Similar developments are visible on a global scale. Mueller (2002, 2010) chronicles various instances across multiple organizations. Denardis (2013, 2014) details similar international tension surrounding the 2012 Dubai World Conference on International Telecommunications (WCIT). It was, however, the Snowden revelations in 2013 that propelled internet governance back into the spotlight (Deibert 2015; Schuster et al. 2017).

The heated debates about these technical topics indicate that the internet is an integral sphere of politics, beyond being a tool for achieving particular technical goals. Second, internet governance is not a technical endeavor that occasionally touches upon politics. Rather, it is inherently political (Denardis 2013, 2014). Some of the most salient questions in the field of internet governance are how such political issues evolve and to what effect (Denardis 2013). This article considers such recent discussions, by focusing on human rights and internet governance.

This paper has five sections: an introduction, methods, introduction to internet governance, a literature review on human rights and internet governance, and a conclusion. Throughout this review, I develop two arguments. First, provide a suggestion on how to more comprehensively approach internet governance. Second, I suggest that current research on internet governance and human rights is well-rounded but skewed towards covering particular technologies, rights, actors, and means for "enabling" human rights online. One of the benefits of such a review is that it highlights academic blind-spots that, given the close inter-relation of internet governance and research, can lead to policy oversights. It is also important to note that such a review is necessarily incomplete and partial, due to various limitations outlined below. Much of the literature on internet governance and human rights aims to answer two questions:

1. Are human rights affected by internet governance, and if so how?
2. Can human rights provide meaningful guidance to internet governance?

While the answers to these questions occasionally overlap, in this paper they will be approached as separate fields of inquiry. This article provides an empirical framing of the literature in emerging natural groupings, which are in line with the taxonomies provided by Van Eeten and Mueller (2013), Raymond and Denardis (2015), and Epstein et al. (2016). However, instead of organizing the literature into distinct categories – as some of the previous reviews on internet governance do – I propose conceptualizing the research using four themes:

1. Technology (content, logical, physical layer)
2. Rights (codified, uncodified)
3. Actors (states, corporations, civil society, etc.)

4. Means (code, law, markets, norms)

The four themes position the subject of human rights in the landscape of internet governance research. The review gives an overview of the different perspectives taken by various academics. The themes are adjacent, non-hierarchical, and occasionally overlap. This approach has the benefit of being rigorous enough to highlight specific issues, while also being sufficiently flexible to indicate where themes overlap. This is necessary for a rapidly developing and often interdisciplinary field. Overall, the literature review highlights novel and well-trodden approaches to indicate what research is done, missing, overemphasized, or not covered in much detail. The literature review can be applied to an array of questions regarding new technologies and human rights.

8.2   Methods

This paper examines the current academic literature on human rights in internet governance, through an integrative literature review (Torraco 2016). Integrative research reviews aim to review, critique, and synthesize 'representative literature on a topic in an integrated way such that new frameworks and perspectives on the topic are generated' (p. 404). The need for such a review is three-fold. First, it enables internet governance research to heed its calls for more diversity (Van Eeten and Mueller 2013). Second, it raises some questions regarding the intellectual relationship between platform governance and internet governance. Third, as the debate on internet governance and human rights is inherently tied to geopolitics (Maréchal 2017), this review suggests additional directions to future research and policymaking.

Overall, for this research more than 130 relevant articles and books were identified. The search strategy for this literature review is broadly based on the method laid out by Torraco (2016 p. 406). This literature review does not claim to be comprehensive. It is important to highlight the limitations of this study. The search terms and databases queried will undoubtedly impact the overall review.[2] Considering the broad approach to internet governance advocated for in this paper, research often not strictly considered within the realm of internet governance, is included. Furthermore, the field is rapidly developing. This work represents preliminary findings from an ongoing doctoral research project and, hence, should be read as work in progress. This literature review does not claim to be definitive; there is ample room for further development. While the body of literature surveyed does not 'tell

---

[2] The following search terms were used: ("internet governance" OR "Internet gov*" OR "cyber law*" OR "telecomm* regulat*" OR "info* sec*" OR "platform stud*" OR "infrastructure stud*") AND ("human right*" OR "digital right*" OR "UDHR" OR "ICCPR" OR "ICESCR" OR "privacy" OR "freedom of expression" OR "political right*" OR "social economic right*"). They were entered into the following data bases: ASSIA, Scopus, IBBS, and Google Scholar.

the entire story' (Torraco 2005 p.361) of internet governance and human rights, it provides various directions forward.

8.3   Internet Governance: A Brief Overview

Internet governance, according to Hofmann (2007 p. 74), 'is a regulative idea in flux'. There are various excellent literature reviews on internet governance (Kahin and Keller 1997; Dutton and Peltu 2005; Mueller 2010; Hofmann et al. 2017). Their understanding is often focused on the technical governance of the internet's core. The reason why many internet governance academics apply this technical definition is to highlight that internet governance, through its technical, multi-stakeholder, private-public nature, is fundamentally different from governance of information and communication through policy and lawmaking (Denardis 2014 p. 19). In this section, I do three things. First, I provide a brief overview of research on internet governance. Second, I argue that there are strong arguments to consider a wider array of technology governance functions in internet governance research. Third, I provide some concrete and theoretical steps for how to integrate these.

By planting focusing on the internet's technical governance, internet governance research runs the risk of overlooking the questions arising out of ongoing cultural, political, and commercial developments. The field might benefit from further considering the dynamic interplay between internet governance narrowly construed and other aspects of technology governance, like internet applications or material hardware. The importance of broadening the definition of internet governance is highlighted by various academics (Mueller 2010; Bygrave 2012), but the question how to do so remains an ongoing topic of discussion. To work towards an answer, it is important to present a brief overview of research on internet governance.

There are multiple definitions of 'internet governance' (Drake 2004; Hofmann et al. 2017). Many academics argue the expression broadly covers the governance of the 'Internet-unique technical architecture rather than the large sphere of information and communication technology design and policy' (Denardis 2014). Internet governance research often describes the particular constellation of actors and organizations that engage in the technical governance of the internet (Mathiason 2008). As well as the importance of coordination between them (Denardis 2010; Hofmann et al. 2017). These actors, actions, and areas of governance are often held up as the distinguishing features of internet governance.

Internet governance and politics are fundamentally intertwined. The current nature of internet governance research cannot be understood without a firm grasp of the political history that shaped it (Mathiason 2008). Following the development of the internet, early research concentrates on governance questions in organizations like the International Telecommunication Union (ITU), the Internet Engineering Task Force (IETF), and ICANN (Mueller 2002; Denardis 2013; Shackelford et al. 2015). The distributed and networked nature of the internet leads much of these discussions to consider its emancipatory (Castells 2000) and economic (Benkler

2006) potential, as well as the role of the state in its governance (Barlow 1996; Goldsmith 1998; Mueller 2002; Goldsmith and Tim 2006).

As the internet became a more pervasive and everyday technology, various academics start to question the narrow focus of the field on a limited number of 'centralized and formalized institutions' (Van Eeten and Mueller 2013 p. 721). They stress the need for more research on 'the diversity of governance on the internet, from centralized, formal global institutions such as the ICANN all the way to the emergent order that arises from the interactions among thousands of ISPs and their users' (p. 730). Such calls for expansion led researchers to cover a broader set of issues while staying within the remit of the technical governance of the internet. For example, Meier-Hahn (2015) looks at how trust and distrust between network operators plays a crucial role in addressing 'the technical, legal and economic uncertainties in the field of Internet interconnection' (p. 1). Mueller and Bendrath (2010) highlight the concerns around deep packet inspection (DPI), free information flows and copyright. Mathew (2016) looks at the Border Gateway Protocol (BGP), which enables separate networks to interconnect, to argue that the internet is a distributed rather than a decentralized system. Much of this research touches directly upon politics, by detailing the economic, social, or political impact of the technology researched as well as the ensuing power struggles to influence governance processes.

The politics of internet design (Braman 2011) and governance (Mueller 2002) are an integral part of internet governance research. Abbate asserts that: 'the debate over network protocols illustrates how standards can be politics by other means' (2000 p. 179). Denardis' 2014 book 'Protocol politics' builds on the work by Abbate and highlights how 'debates over protocols bring[ing] to light unspoken conflicts of interest' (p.10). While her work focuses mostly on internet protocols, Denardis does highlight that 'politics are not external to technical architecture' (p.10). This argument can be extended to include a broader set of questions on the governance of the internet's design, in internet governance research.

Upholding this understanding of internet governance, while also arguing that it is 'politics by other means' (Abbate 2000 p. 179), makes it difficult to justify why only the 'technical governance' of the internet can be considered internet governance. Or how it can be known what constitutes 'technical governance'. If internet governance is inherently political, then it could also further consider other internet design governance questions.

Recent work on social media platforms design (Gillespie 2010; Denardis and Hackl 2016; Klonick forthcoming), or web traffic architecture (McIlwain 2016), and the internet's physical hardware, like deep sea cables (Ross 2014; Starosielski 2015), suggest that such a broadened focus will open up interesting new research paths. Rather than calling for more research that views these questions through 'an Internet governance lens' (Denardis and Hackl 2016 p. 762), it is interesting to approach them as an integral to internet governance. However, this does not mean that every single question of internet content, use, or hardware needs to be included in internet governance research. Rather, it means including questions about the governance of the internet's design, broadly interpreted. Suggesting this approach, without indicating how to demarcate what falls within its remit and what does not, can lead to confusion. However, as the internet and its related technologies expand constantly it is important to maintain a flexible approach. This broadened approach can, however, be bound by two properties: first, it concerns the design of the internet as it pertains to not only to protocols, but also to Internet platforms and physical pipes. Second, it considers how that design is governed. This approach can undoubtedly be further refined, and perhaps this paper provides another step towards that goal.

Broadening the approach to internet governance research in this way is relevant because it ensures that research stays at pace with the changing role of the internet in society. This broader approach captures not only the importance but also the intricacies of the governance of internet design decisions. It also shows how a wide set of design decisions impact the technical governance of the internet (Internet Architecture Board 2018),[3] narrowly defined. For instance, within the Internet Engineering Task Force (IETF) various information intermediaries – like Google – are successfully developing standards based on their in-house designs. These standards can benefit the technical functioning of the internet as a whole, but they also facilitate the information intermediaries' business models. Just like Helmond (2015) shows how the economic models of online platforms are shaping the design of the world wide web, so must internet governance research look at how such information intermediary design dynamics shape internet governance. This is particularly important considering that these information intermediaries directly impact society. Or as MacKinnon holds: 'Unlike companies that produce sportswear or toothpaste, the value proposition of Internet related companies relates directly to the empowerment of citizens' (2012 p. 172).

Considering a broader understanding of internet governance brings two specific implications for academic scholarship on this topic. Most notably, it means widening the disciplinary and methodological scope of internet governance research. In addition to arguing for a broader definition, it is also important to consider what disciplines are represented (Van Eeten and Mueller 2013). Internet governance is an inherently interdisciplinary field (Hofmann et al. 2017). But particular disciplines, for example, computer science, sociology, law, and economics (Van Eeten and Mueller 2013; Epstein et al. 2016) are better represented than others, such as anthropology and cultural studies. This also means that some of the crucial questions surrounding culture (Abbate 2000), human agency (Van Eeten and Mueller 2013), social ordering (Flyverbom 2010; Hoffman et al. 2017), information intermediaries (Denardis and Hackl 2016), and hardware (Zittrain 2008) remain unanswered. Even though internet governance research increasingly takes on a Science and Technology Studies (STS) perspective (Flyverbom 2010; Brousseau et al. 2012; Denardis 2014; Musiani et al. 2015), it needs to commit to bringing in a wider set of fields, especially those from disciplines like cultural studies, critical race studies, and anthropology.

Recent work from these disciplines indicates, for example, that the overarching themes of gender, race, and identity politics are both relevant to, and underexplored

---

[3] Internet Architecture Board. (2018). Consolidation. Retrieved from https://www.ietf.org/blog/consolidation/

in, internet governance. McIlwain (2016), for instance, shows how racial inequality is (re)created through the architecture of the web, yet there is limited understanding of how this happens in the governance of the technical and material architecture of the internet. Likewise, the work by boyd (2012) shows that race and class have an impact on social networking amongst teens, but there is limited knowledge about such racial and class-based dynamics in internet governance. Similarly, the work of Abbate (2012) and Hicks (2017) details the hidden history of women in computing and gender discrimination, which is likely to have parallels in internet governance. These questions are relevant because the internet isn't magic. It is fundamentally human (Abbate 2000) and, as such, its governance is unlikely to be immune from (identity) politics.

Second, bringing in a broader set of academic disciplines allows internet governance research to further explore the material aspects of the internet. There is a shift towards the 'materiality' of the internet within the academic fields of media and communication studies and STS (Dourish 2015). This shift is also finding sustained academic attention in internet governance (Musiani et al. 2015; Epstein et al. 2016). It is important to further develop it because everyday material life – from light bulbs to toothbrushes to cars – is increasingly connected to the internet (Kopetz 2011). To understand these dynamics, it is vital to bring in people with the right methodological tools to expose them.

Ziewitz and Pentzold (2014) argue that more attention should be paid not just to defining internet governance, but also to how these definitions create internet governance research as much as they describe it. This is a crucial insight, as it ties together what this first section does: it presented an overview of internet governance research, suggested broadening its remit and, highlighted the necessity of including additional theoretical and methodological approaches. This is particularly relevant when positioning the subject of human rights in internet governance research.

8.4   Positioning Human Rights in Internet Governance

As was mentioned in the methods' section, it is important to highlight and understand the role of human rights in internet governance for academic and policy purposes. Internet governance, even in its most narrow definition, has an impact on public interest issues like human rights (Solum 2009). Even though policymakers generally pay attention to the importance of having human rights guide internet governance (Jørgensen 2013), few broad-scope literature reviews, positioning the relevant research on human rights in internet governance research, exist. This literature review takes the broad understanding of internet governance, as outlined above, and includes research on the governance of the internet's content, logical, and material layers (Benkler 2000).

Understanding how human rights are discussed in internet governance is important because various internet governance organizations are explicitly incorporating human rights norms in their governance structures. For instance, ICANN's new bylaws include a commitment to respect 'internationally recognized human rights as required by applicable law'(ICANN 2016).[4] The Global Network Initiative (GNI),[5] a coalition of industry, civil society and academia, developed a set of principles based on human rights (Baumann-Pauly et al. 2017), to resist pressure by governments to comply with domestic laws that are irreconcilable with the United Nations (UN) Universal Declaration of Human Rights (UDHR). The Human Rights Protocol Considerations Research Group (HRPC-RG) in the Internet Research Task Force (IRTF), the IETF's parallel research organization, is developing human rights protocol considerations for IETF engineers.[6] Large software and hardware companies like Microsoft and internet access providers like dot.NL registry 'Stichting Internet Domeinregistratie Nederland' (SIDN) (Simon 2018)[7] are undertaking human rights impact assessments (Microsoft 2017).[8]

This review is also relevant because human rights are not uncontested (Moyn 2012). While often presented as carrying an inherently emancipatory agenda, it is important to carefully highlight what is meant by human rights,[9] which actors are advocating for it, using what technology, and means to achieve which ends (Cath and Floridi 2017). The literature review suggests that there are four (preliminary) themes of research on human rights in internet governance: Technology, Rights, Actors, and Means. Often, research is categorized as belonging to more than one theme. This review allows the research to be placed relative to the various themes it explores.

8.4.1   Technology

The first theme is the technology proper. Without understanding what internet technology is discussed and how it is governed, any discussion of human rights would be impossible. To understand the theme of technology, it makes sense to use a taxonomy that conceptualizes the internet as existing out of various layers. The previous section detailed how much research on internet governance is geared towards the technology residing at the 'logical' layer. It indicated that researchers taking

---

[4] ICANN. (2016). Bylaws for Internet Corporation for Assigned Names and Numbers. A California Nonprofit Public-Benefit Corporation. Retrieved from https://www.icann.org/en/system/files/files/adopted-bylaws-27may16-en.pdf

[5] Global Network Initiative. (2017). The GNI Principles. Retrieved from https://globalnetworkinitiative.org/gni-principles/

[6] Disclosure: The author of this paper is a co-author on this work.

[7] Simon, Maarten. (2018). Sure, we respect human rights. Don't we?. Retrieved from https://www.sidn.nl/a/about-sidn/sure-we-respect-human-rights-dont-we

[8] Microsoft. (2017). Human Rights. Retrieved from https://www.microsoft.com/en-us/about/corporate-responsibility/human-rights

[9] This will be done in sect. 4.2

technology as their academic starting point tend to focus on the 'logical' (protocols) and 'content' (platforms) layer, missing the internet's material technology (pipes).

Many different taxonomies exist for understanding the various layers of the internet. These taxonomies can be split up into having technical and non-technical origins. They serve different purposes. It is important to understand when they are used in internet governance research (Bygrave 2012) as this informs how human rights are approached. There are three well-known examples of internet layer taxonomies that have non-technical origins. The model by Benkler (2000 p. 562), for example, divides the internet up into three layers: the content layer (information to interact with), the logical infrastructure layer (technical infrastructure, like software and protocols), physical infrastructure (layer-wires, cable, radio frequency spectrum). Benkler developed this model to understand media regulation in the burgeoning digital age and direct it such that the internet would allow for new forms of decentralized communication.

Lessig built on this model by transforming it to explain what options exist for influencing behavior on the internet (2006). His idea – often referred to as the 'pathetic dot theory' – is widely used and as such will be covered in Sect. 4.4. The four-layered model by Choucri and Clark (2012 p.2) is also based on the work of Benkler. Their multidisciplinary approach divides the internet up into a social, information, logical, and physical layer.[10] This model aims to bring together key concepts in international relations with those of 'cyberspace'. It provides a granular understanding of how the levels of analysis common in international relations: individuals, states, and the international system, apply to cyberspace. This research follows the taxonomy provided by Benkler, as it most closely aligns with the themes of the review. In the next section, I give an overview of the research on human rights at the internet's content, logical, and physical infrastructure layer.

8.4.1.1 The Content Layer

There is a growing body of literature looking at the governance of the internet's content layer and human rights through platforms. For instance, Gillespie (2010) looks at how the term platform obfuscates the extent to which social media companies actively intervene through design in content management, posing challenges to the right to freedom of expression. Helmond (2015) argues that platforms are the main infrastructure and economic model for the social web, and that their power is leading them to drive the overall web to be geared towards commercial needs. Denardis and Hackl (2016) look at how platforms constrain various human rights and what role their business models play in this. They make a strong case for why internet governance as a field should include more research on the content layer.

---

[10] The social layer referring to the individuals on the internet, information to the data on the internet (social media posts, blogs), the logical layer includes the digital infrastructure of the internet (the protocols and standards that enable interoperability), the physical layer covers all of the tangible aspects of the internet (deep sea cables, laptops).

Almeida et al. (2016) look at how app stores fit into the internet governance ecosystem. They argue that the design of app stores undermines users' rights (p. 50). Jørgensen (2017) shows that the individuals working at Facebook and Google understand human rights mostly in terms of state violations instead of acknowledging the impact their corporate design decisions have. Lucchi (2014) discusses internet content governance and argues that human rights can provide an 'institutional safeguard against the expansionary tendency of market powers and on the increasing role of the courts in expanding and adapting the frontiers of fundamental legal rights.' (p. 845). This literature supports the assertion made earlier about the advantages of expanding internet governance research to include content governance. It also suggests that the governance of design at this layer has implications for human rights and plays into design governance questions at the logical layer.

8.4.1.2   The Logical Layer

Much of the research on human rights and internet governance is focused on the logical layer (Mueller 2002; Denardis 2014; Musiani et al. 2015). This can plausibly be explained by the historical development of internet governance research. Rundle and Birdling (2008), for instance, detail how internet filtering at the logical layer obstructs human rights and argue that international human rights standards are the proper framework for setting the limits of reasonable use of internet filtering. Cath and Floridi (2017) show how engineers at the IETF understand their responsibility to respect human rights through protocol and standards.

Brown et al. (2010) argue that universal human rights values should be baked into the design of the Internet. Rachovitsa (2016) highlights the work of the IETF on privacy considerations as an example of how privacy-by-design can be achieved at the logical layer. Research on human rights and internet governance at the logical layer often focuses on many of the larger Internet governance organizations. There is room for research on smaller regional actors, like Internet Service Providers (ISPs) (Brown et al. 2010).

8.4.1.3   The Physical Layer

Research on internet governance at the physical layer is growing. Examples include interdisciplinary research on undersea cables (Starosielski 2015) and the infrastructure of the internet in New York (Burrington 2016). What this type of research does is show how the governance of physical layer is tied up in politics, culture, and the material features of the landscape in which it is embedded. Although rarely explored as such, this layer is bound to have implications for human rights (Ross 2014), as well as pose new questions regarding the usefulness of international human right standards as guiding frameworks for internet governance. These questions are surfacing in international politics (Policy Exchange 2017)[11] but not often translated to academic research.

8.4.1.4   Conclusions

Concluding, much of the current research on internet governance and human rights focuses on the logical layer. There is, however, a growing body of literature across multiple disciplines that looks at questions of internet content governance and human rights. There is limited research on the physical layer, even though it is likely that there are relevant human rights' questions in this area. Overall, the existing literature primarily describes how human rights are impacted by internet governance. However, questions surrounding the extent to which human rights can provide prescriptive guidance, is gaining traction as the next paragraphs show.

8.4.2   Rights

The second theme of research takes as its departure point different human rights. This review follows the definitions of human rights as reflected in the existing literature. As such, it extends beyond the legal understanding of human rights encoded in international frameworks, like the Universal Declaration of Human Rights (UDHR), the International Covenant on Civil and Political Rights (ICCPR) and the International Covenant on Economic, Social and Cultural Rights (ICESCR), and in national frameworks. In this section, I provide an overview of how ongoing research defines and approaches human rights and show what gaps exist.

Much of the initial research conceptualized human rights as per the international human rights' frameworks, approaching human rights as indivisible, interrelated, and interdependent (Jørgensen and Marzouki 2015). Later research zoomed in on political rights like freedom of expression and privacy (Padovani et al. 2010). Simultaneously, various suggestions for additional rights, currently uncodified, surged. Examples are the 'right' to Internet access (Joyce 2015) and the 'right' to network self-determination (Belli 2017).[12] The literature indicates that there is a need to further evaluate to what extent existing human rights' frameworks can provide legal and normative guidance to internet governance.

---

[11] Policy Exchange. (2017). Undersea Cables: Indispensable, insecure. Retrieved from https://policyexchange.org.uk/publication/undersea-cables-indispensable-insecure/

[12] Belli, Luca. (2017). The scramble for data and the need for network self-determination. Retrieved from   https://www.opendemocracy.net/luca-belli/scramble-for-data-and-need-for-network-self-determination

8.4.2.1   International Human Rights Framework: Indivisible, Interrelated, and Interdependent

Initial research on human rights and internet governance builds on previous academic and policy work done on the 'information society' (Metzl 1996; Hick et al. 2000). It covers a broad set of human rights frameworks, approaching human rights as fundamentally indivisible, interrelated, and interdependent. The impact of internet governance on human rights came to the fore on a global scale during the World Summit on the Information Society (WSIS) in 2003 and 2005. During the second phase of this UN convened summit the 'Tunis Agenda for the Information Society' (ITU 2005)[13] was drafted (Mathiason 2008 p. 97). This statement was one of the first instances linking human rights[14] and the internet in a UN document.

Multiple impactful academic works on internet governance and human rights were published following the WSIS conferences. The edited volume by Drake and Wilson (2004) covers a variety of internet governance issues, highlighting the perspective of less influential or 'non-dominant' actors. The various topics covered reflect ongoing political disagreements regarding internet governance and human rights. For instance, the tensions between the EU and US approach to the right to privacy as well as property right disputes between Global North and Global South regions. The edited volume by Jørgensen (2006) provides an interdisciplinary overview of the various issues surrounding human rights, politics, and the internet. The volume explicitly covers a broad spectrum of political and socio-economic human rights. Various authors in these volumes are policymakers or activists as well as academics, suggesting that academic work in this area is often informed by a practitioner's perspective.

This is also visible in contemporary academic debates. In 2012, the UN Human Rights Council (OHCHR) adopted a resolution that affirmed that 'people's rights offline must also be protected online' (OHCHR 2016).[15] This resolution, like other UN documents, suggests that the Internet is a fundamental enabler of human rights and implies that the current understanding of human rights, as laid out in the UDHR, can be applied to the internet (Jørgensen 2006). According to this resolution, there

---

[13] ITU. (2005). Tunis Agenda for the Information Society. Retrieved from http://www.itu.int/net/wsis/docs2/tunis/off/6rev1.html

[14] Article 42 of the Tunis Agenda holds: '42. We reaffirm our commitment to the freedom to seek, receive, impart and use information, in particular, for the creation, accumulation, and dissemination of knowledge. We affirm that measures undertaken to ensure Internet stability and security, to fight cybercrime and to counter spam, must protect and respect the provisions for privacy and freedom of expression as contained in the relevant parts of the Universal Declaration of Human Rights and the Geneva Declaration of Principles.' See http://www.itu.int/net/wsis/docs2/tunis/off/6rev1.html The Tunis agenda also led to the creation of the Internet Governance Forum (IGF), an annual UN-convened multistakeholder meeting on internet governance without a concrete policy-making mandate where stakeholders can meet to discuss emerging issues.

[15] UN Human Rights Council. (2016). Resolution on the promotion, protection and enjoyment of human rights on the Internet (A/HRC/32/L.20). Retrieved from http://ap.ohchr.org/documents/dpage_e.aspx?si=A/HRC/32/L.20

is no need to develop new human rights or tailor the existing ones to the digital age (Carr 2013). Much of the initial academic work done on human rights and internet governance follows this policy line, but there is a growing body of work that critically assesses the applicability of human rights frameworks to internet governance (Land 2013; Carr 2013; Orwat and Bless 2016; Mueller and Badiei 2018).

More research is needed to understand whether codified human rights frameworks can provide legal and ethical guidance to internet governance (Klang and Murray 2004; Easton 2016). Especially, as the premises underpinning international human rights standards may be affected by the transformation brought about in societies by the internet (Mathiesen 2014). Related, there are many outstanding questions regarding the utility of these frameworks, especially when considering private governance, in addressing broader social justice questions (Lettinga and Van Troost 2015).

8.4.2.2 International Human Rights Framework: Political Rights

A second sub-theme of research on rights arises with the mainstream adoption of the internet. It prioritizes research on political rights, like freedom of expression and privacy (Klang and Murray 2004; Brousseau et al. 2012; Land 2013; Jørgensen and Marzouki 2015). This shift in focus, from a broad array of human rights to a narrower set, can plausibly be explained by the changing role of the internet: it went from an academic network to an everyday medium for retrieving, sharing, and storing information (Naughton 1999). Related, internet governance research increasingly looks at copyright questions (Mueller 2002; Denardis 2013). Copyright questions, in turn, have a direct bearing on political rights like freedom of expression according to prominent scholars (Lessig 2004).

As the internet continues to 'envelop' (Floridi 2014) every aspect of society, its governance impacts a broader array of human rights, including socio-economic rights (Marsden 2008). For example, the governance decisions of online labor platforms influence the right to work (Graham et al. 2017). Likewise, scholars argue that Artificial Intelligence (AI) systems can impact the right to non-discrimination (Noble 2018) and equality (Eubanks 2018). Recent research starts to bring in the human rights issues of minorities in internet governance, like the LGBTQIA community (Hackl 2016) and linguistic minorities (Moller 2013), as well as perspectives from the Global South (Santoro and Borges 2017). As the Internet continues to change and expand, it is important for academics to further explore such questions.

8.4.2.3 Beyond the International Human Rights Framework: New Rights

The literature surveyed surfaced a third sub-theme, centered on the question whether there is a need for new human rights.[16] The overall discussion ranges from making access to the internet a human right, to the 'right' to be forgotten, to the right to

---

[16] Examples of this are discussions about the 'right to be forgotten' or the 'right to Internet access'. Neither of which are recognized within the international framework for human rights but are approached in both academic and popular discourse using a human rights framing.

network self-determination (Belli 2017),[17] to creating a digital Geneva Convention (Smith 2017),[18] to developing an 'Internet Bill of Rights' (Musiani 2012). It also includes discussions on establishing an international legally binding instrument on 'Transnational Corporations and Other Business Enterprises' with respect to human rights (Ruggie 2017) and whether human rights should be applied to groups in addition to individuals (Taylor et al. 2017). Some academics argue that such new rights can have negative outcomes (McGoldrick 2013; Jørgensen and Marzouki 2015; Alston 2017), but others hold it is important to consider new mechanisms to ensure the internet has a net-positive impact on society (Redeker et al. 2018).

One of the most controversial debates within this sub-theme is whether access to the internet should be a human right (Karanicolas 2013; Tully 2014). The initial idea was spearheaded by civil society just after the second WSIS conference in 2005. Since then, the debate continued with famous internet pioneers like Tim Berners-Lee (2014)[19] and Vint Cerf (2012)[20] taking opposing sides. Although interesting, the focus on this sub-theme can lead researchers to overlook equally relevant questions around socio-economic rights and human rights as a guiding framework for internet governance mentioned earlier.

8.4.2.4 Conclusions

This section looked at the theme of human rights. It indicated that there are three sub-themes in the research: a broad focus on codified rights' frameworks, a narrow focus on codified political rights, and new rights. There is much academic debate about the role of human rights in internet governance, but the different approaches to what is meant by the terms can lead academics to talk at cross purposes. This means that academics doing related work do not necessarily understand their work as sharing a common denominator: internet governance. This section also showed that there is room for further research on socio-economic rights, minority rights' issues, and the suitability of international and national human rights frameworks as guiding principles for internet governance.

---

[17] Belli, Luca. (2017). The scramble for data and the need for network self-determination. Retrieved from https://www.opendemocracy.net/luca-belli/scramble-for-data-and-need-for-network-self-determination

[18] Smith, Brad. (2017). The Need for a Digital Geneva Convention. Retrieved from https://blogs.microsoft.com/on-the-issues/2017/02/14/need-digital-geneva-convention/

[19] Berners-Lee, Tim. (2014). Recognise the Internet as a human right, says Sir Tim Berners-Lee as he launches annual Web Index. Retrieved from https://webfoundation.org/2014/12/recognise-the-internet-as-a-human-right-says-sir-tim-berners-lee-as-he-launches-annual-web-index/

[20] Cerf, Vint. (2012). Internet Access is not a Human Right. Retrieved from http://www.nytimes.com/2012/01/05/opinion/internet-access-is-not-a-human-right.html

8.4.3   Actors

The third theme concerns the actors discussed in the context of internet governance and human rights. This research theme is often closely tied to geopolitical developments.[21] Much of the research mirrors the developments in internet governance mentioned in sect. 3: it focusses on large institutions (Badii and Mueller 2018; Cath and Floridi 2017) as well as state actors (Carr 2013). The institutions, in this case, are not just technical organizations like ICANN and the IETF, but also content layer actors like Facebook and Twitter (Land 2013; Hackl 2016; Almeida et al. 2016; Jørgensen 2013, 2017; Balkin forthcoming; Klonick forthcoming).

The focus on these content actors can be explained by their role in mediating the everyday use of the internet, as well as the recent controversies around their impact on election outcomes. Considering the scale and network effect of these companies, the implicated human rights issues are highly relevant. However, it is important to further develop knowledge of the more 'invisible' technical private actors. Interestingly, academics focusing on the 'technology' theme do look at these invisible technical actors but tend to overlook content layer actors.

Part of this disconnect can be explained by what disciplines are well represented in these various thematic discussions. Where 'classic' internet governance academics naturally migrate towards questions about the governance of technology at the logical layer, others like media and communication academics focus on platforms.

8.4.3.1   Non-State Actors

The influence of non-state actors, frames many of the academic debates about internet governance and human rights (MacKinnon 2012; Land and Aronson 2018; Jørgensen forthcoming). The literature surveyed highlights various strands of research on non-state actors: the actors maintaining the infrastructure through which internet content flows, the corporate actors creating this content or shaping the platform design on which it's produced, and the civil society actors advocating for human rights.

This type of research is particularly relevant as various non-state actors are dealing with human rights questions internally. Braman (2017) documents how 'the computer scientists and electrical engineers involved in the internet design process dealt with social policy and political issues in the course of their design work' (p.70). Her work shows that engineers consider, in careful detail, various human rights questions, like privacy, from the network's beginnings. Debunking common assumptions about engineers' disinterest in such issues. Jørgensen (2017) demonstrates that how

---

[21] Some notable examples are the World Summit on the Information Society (WSIS) conferences in 2003 and 2005; the 2012 UN resolution that held that human rights apply online as they do offline; the Snowden revelations in 2013 that revealed the extent of US communications-intercept operations using the internet's physical and logical layer; and the subsequent Global Multistakeholder Meeting on the Future of Internet Governance (NETmundial) in 2014.

internet platforms understand human rights is tied to their business models. She also shows that this understanding impacts the design features of platforms, which ties into questions of internet governance. Suzor et al. (2018) outline how human rights values can be used to create an index that rates the legitimacy of the governance undertaken by online intermediaries. Interestingly, they define intermediaries broadly making their suggestions relevant to an array of internet governance actors.

Even though internet governance research always considered the role of civil society (Mueller 2002; Mathiason 2008), there are many outstanding questions regarding their role in steering the recent uptake of human rights (Appelman 2016; Milan and Ten Oever 2016; Cong 2017). The last decades saw the growth of international human rights NGOs and an expansion of representation of civil society concerns in internet governance bodies (Benedek 2008; Bygrave and Bing 2009; Hintz and Milan 2009), for instance through the work of the Non-Commercial Users Constituency (NCUC) in ICANN (Mueller 2002). But there is little comprehensive understanding of how including human rights in internet governance impacts the ability of the various actors, including civil society, to shape information flows (Mueller 2016).[22] It undoubtedly has an impact. Research show that the inclusion of human rights can reaffirm existing arrangements of power and increase private ordering of laws and information control (Scholte forthcoming), but it can also lead to the development of novel mechanisms for private arbitration of public values (Vezzani 2014).

8.4.3.2 State Actors

States are the principal entities responsible for upholding human rights. As such, specific exploration of what this responsibility could and should look like in 'information societies' (Floridi 2017) is necessary. The cyber-utopian views of the early nineties (Barlow 1996) – which predicted the internet would remain fundamentally free from government influence, regulation or territorial law (Johnson and Post 1997) – are argued to be untrue (Brown and Marsden 2013) and undesirable (Lessig 2006). But there remains a contentious debate about the role of states in internet governance (Drezner 2004; Benkler 2006; Goldsmith and Tim 2006; Christou and Simpson 2009; Brown and Marsden 2013; Denardis 2013).

The ability of states to protect or violate human rights through internet governance is in flux (Carr 2015; Kiggins 2015; Scholte forthcoming). Deibert et al. (2008) highlight that human rights violations, such as privacy and freedom of expression but also children's rights, increasingly take place through manipulation of the internet's infrastructure by private actors. Sometimes by their volition, other times at the behest of states. Maréchal (2017) looks at how states use the internet for geopolitical purposes such as election manipulation. Mueller discusses content gov-

---

[22] Mueller, Milton. (2016). Missing the Target: The Human Rights Push in ICANN goes off the rails. Retrieved from https://www.internetgovernance.org/2016/10/26/missing-the-target-the-human-rights-push-in-icann-goes-off-the-rails/

ernance and suggests a two-tiered system, the second being geared towards account-ability 'in which governments intervene to counteract excessive market power or when basic human rights are violated' (2002 p. 213). Lucchi (2014) argues that access to the internet is a fundamental requirement to enable human rights in a con-nected world and suggests steps states should take to increase access. Overall, the literature shows that there are many open questions about the role of states at the intersection of internet governance and human rights (Van Hoboken 2012).

8.4.3.3   Conclusions

This theme indicated that current research covers both state and non-state actors. It highlights how research themes sometimes overlap, covering questions of rights, technology, and actors. This often happens when academics take a multi-disciplinary approach. It also showed that there is much attention on how the actors at the con-tent layer approach human rights and translate them into their design governance decisions. Similar research is needed covering less visible technical actors, espe-cially those that are explicitly incorporating human rights considerations into their work. There is also a need for more research on the role of civil society in steering the uptake of human rights in internet governance. Such research will shape aca-demic understanding of the strategic interests and specific engagement of the vari-ous actors leading the debate on, and the uptake of, human rights in internet governance. Likewise, the ability and legitimacy of non-state actors, like technical and corporate organizations, to take on human rights concerns is not thoroughly explored. Especially, as human rights frameworks are premised on the state being the upholder of these rights.

8.4.4   Means

A fourth and last theme is the means through which the exercise of human rights can be 'enabled' or 'mediated' through internet governance. These various means also influence how human rights are impacted by internet governance, the most promi-nent such examples are covered in the previous sections. Lessig (2006) identifies four means of regulating behavior: code (or architecture), norms, laws, and markets. As internet governance is influenced by these four fields of social control, the litera-ture on how human rights can be enabled will be classified using Lessig's theory.

8.4.4.1   Code

Regarding code, Lessig, famously argues that the internet's architecture can enable but also disable the exercise of human rights. In that same vein, Denardis (2012) discusses various points of control within the internet's architecture. She argues that internet governance, through its design of code, mediates human rights' affordances. Overall, the literature suggests that the architecture of the internet enables (Lessig 2006), mediates (Denardis 2014), and structures (Musiani et al. 2015) the exercise of human rights.

Various academics argue that there is a need to experiment with new, technical, means beyond strict legal compliance (Land 2013; Rachovitsa 2016; Alston 2017) for enabling human rights through internet governance. Reidenberg (1997) holds that such technical approaches address the various problems not solved by legal strategies. Orwat and Bless (2016) contend that direct instantiating of rights in internet protocols can work in some cases, but not all. Land calls on 'technology companies to embed 'human rights defaults' into their technology by designing it in ways that make it harder for states to violate international human rights' (2013 p. 2). But how to do so specifically remains unclear. Much of the current literature on human rights and code builds on the academic work done on embedding values 'by-design' (Wiener 1954; Cavoukian 2012; Friedman et al. 2013). Yet, this approach is not without its critics.

There are various obvious concerns. First, the lack of a clear methodology for achieving values-by-design remains an issue. This is particularly thorny in the case of human rights, as there are inherent tensions between human rights. Second, even if these issues are addressed there is still the problem related to the lack of business incentives for including values-by-design (Bygrave 2017). Mueller and Badiei (2018) argue that human rights cannot be enabled through code and attempts to do so are misguided because they fundamentally misconstrue the complexities of the interaction between technology and society. There is a need for further exploration of how the well-known drawbacks of the 'by-design approach' play out when discussing internet governance and human rights.

8.4.4.2   Laws

Some of the relevant debates on the promises and perils of using laws to promulgate human rights through internet governance are covered elsewhere in this article. These include the lack of binding human rights obligations for non-state actors in the face of a privately-owned internet (Laidlaw 2012; Klonick forthcoming; Jørgensen forthcoming) and the tensions between national legal requirements and international law (Lucchi 2014).

Many researchers think about (the limitations of) enabling or mediating human rights through code, however additional research that specifically looks at how existing or new human rights laws can be applied to internet governance is needed (Suzor et al. 2018). There is some research on how laws could be used towards enabling human rights in internet governance. Brown and Marsden (2013), for example, call for 'prosumer law' to protect human rights by ensuring that 'code' is not just driven by commercial interests. Land (2013) argues that international regulation of the internet – and its governance – must include attention to existing international human rights law. However, other academics hold that it is crucial to be aware of the auxiliary effects, of such human rights' law-based approaches to internet governance, on reinforcing state power (Antonova 2013; Carr 2013) and weakening media freedom (Lucchi 2014).

Related, there is a growing body of literature on digital constitutionalism, which seeks 'to advance a relatively comprehensive set of rights, principles, and governance norms for the internet, and are usefully understood as part of a broader, proto-constitutional discourse' (Gill et al. 2015). Some authors argue that this line of research is developing separate from internet governance (Pettrachin 2018). Others place it squarely within internet governance (Kulesza 2008; Suzor et al. 2018). The debate on digital constitutionalism is particularly interesting because it highlights the need to focus on an array of socio-economic rights as well as new rights.

8.4.4.3   Norms

Some of the current work on internet governance and human rights considers norm setting through policy forums, like the Internet Governance Forum (IGF) (IPRC 2018).[23] Antonova (2013) tracks how IGF debates establish 'the existing international human-rights platform as the defining principles of Internet governance' (p. 93). She highlights five obstacles: the layered nature of the internet, the state-oriented nature of the human rights framework, the lack of specificity in terms of what issues bringing in the international human rights framework addresses, and the lack of a clear goal towards which the human rights framework is evoked. These issues notwithstanding, in her book 'Consent of the Networked' Mackinnon (2012) argues that human rights can provide explicit norms for internet governance. Similarly, Orwat and Bless (2016) suggest that 'for the Internet as a global communication infrastructure, the relevant catalog of social values are human rights' (p. 25). Mueller et al. (2007) go into more detail and argue in favor of establishing a framework convention on internet governance defining principles and norms rooted in international law. While many support the call for human rights norms (Maréchal 2015), additional research is needed on the difficulties surrounding the operationalization of such calls.

Similarly, many official publications coming out of international or regional organizations like the European Union, the United Nations General Assembly, and the United Nations Human Rights Council also advance norm setting in internet governance using the international human rights framework. At the same time, when discussing the governance of emerging technologies like AI, a turn by governments away from hard regulation to softer norm development based on 'ethics' is visible. This is relevant as it shows a change in approach by policymakers, which is likely to echo in research on norm setting for internet governance.

---

[23] Internet Rights and Principles Coalition. (2018). IPRC Charter. Retrieved from http://internetrightsandprinciples.org/site/charter/ The IRPC is an example of such ongoing work on norm setting.

8.4.4.4 Markets

Markets play a crucial role in shaping internet governance (Mueller 2002). Much of the research looks at how private actors are driven by their business models which subsequently inform their human rights' ramifications. Lessig originally meant pricing constraints when discussing market forces online (2006 p. 139). But with regards to human rights and internet governance, relevant questions are also around digital convergence (Mueller 2010 p. 9), market monopolies, network effects (Nielsen and Ganter 2017) and the data-driven economy (Williams forthcoming).

The literature on markets and human rights covers both questions of logical and content layer governance. Powers and Jablonski (2015) argue that the development of the internet is ultimately aimed at advancing Western economic and political interests, as opposed to human rights. Bygrave presciently argues that when the internet enables behavior that presents a danger to human rights, its reliance on contracts might change the relative role played by contract and statute, favoring the latter over the former (2012 p.20). Aaronson (2015) looks how the US and the EU try to advance human rights values through trade-agreements, and where these attempts fall short. All of this work suggests that the constantly shifting landscape of internet governance and the impact of its market dynamics on enabling human rights warrant more academic attention, especially as trade organizations and agreements are becoming further implicated in internet governance.

8.4.4.5 Conclusions

Current research on how human rights can be enabled through internet governance can be classified using Lessig's four means of control: code (or architecture), norms, laws, and markets. There is lively academic discussion on all four means. In particular surrounding the question of the feasibility of using code for enabling the exercise of human rights. Most authors agree that invoking one modality to achieve the desired outcome is unlikely to work. It takes a combination of forces for internet governance to enable human rights in a meaningful way. The statements on the ability of various means to enable human rights through internet governance needs further empirical grounding.

8.5   Conclusions

The main contribution of this literature review is twofold, first I provided a suggestion for how and why to consider a broader array of technology governance and design decisions as part of current research on internet governance. Through its political nature, internet governance is inherently influenced by and connected to wider questions of platforms and physical pipe governance.

The driving motivation for including a broader set of questions in internet governance research is because it enables the research to stay on par with commercial, political, cultural, and social developments that shape the role of the internet in society. It does not, however, mean that all issues related internet content, use, or hardware should be included in internet governance research. Rather, more work on questions about what constitutes the governance of the internet – its epistemology – is needed (Cath forthcoming). This means, for example, understanding how the governance of content and material layer fit into the overall research agenda. Consideration of how to expand current approaches to internet governance research is also important because it brings disparate academic communities in closer conversation.

Second, I presented an integrative literature review positioning human rights in internet governance. The review synthesizes the surveyed literature into natural groupings, that are presented in the form of four themes: technology, rights, actors, means. It recasts internet governance to include human rights questions in content and physical internet design governance. The review supports researchers to both situate and design their future research projects, as well as access relevant findings from other disciplines. It highlights gaps in the literature and the bias towards particular research. For example, research on technology often primarily looks at the logical layer. Research on human rights is chiefly geared towards political rights and new rights, overlooking socio-economic rights. Research looking at actors mainly covers questions of content governance. Research on means for enabling human rights focuses mostly on code.

The overall coverage of human rights in internet governance is well-rounded. However, there are clear gaps. Which arise, in part, because researchers focusing on specific technology, rights, actors, or layers do not always consider research done in other disciplines to be relevant, as it does not align with their 'version of governance' (Ziewitz and Pentzold 2014 p. 307) or who they consider to be part of their epistemic community.

The review also shows that advancing human rights through internet governance is not a neutral endeavor. The internet, as a global network, can benefit from being built following these socio-legal standards. It may ensure consistency across time and space and add a degree of democratic and normative legitimacy to design decisions. However, while some argue that it is both ethically desirable and legally justifiable to do so, it is important to understand how human rights are operationalized and by extension whose agenda is being furthered. It is also important to consider what alternative agendas or frameworks, like those based on social justice or ethic, can structure internet governance such that it upholds public interest values.

Acknowledgements   I would like to thank the two editors for their helpful feedback and comments. I am also thankful for the feedback I received from Professor Luciano Floridi and Dr. Victoria Nash, as well as various members of the Digital Ethics Lab (DELab). I also want to extend a special thanks to Vidushi Marda, Niels ten Oever, Christiaan van Veen, Ben Zevenbergen, and Jo Gardner.

References

Aaronson, Susan. 2015. Why Trade Agreements Are Not Setting Information Free: The Lost History and Reinvigorated Debate over Cross-Border Data Flows. *Human Rights, and National Security* 14 (4): 671–700. https://doi.org/10.1017/S1474745615000014.

Abbate, Janet. 2000. *Inventing the Internet*. Cambridge MA: MIT Press.

———. 2012. *Recoding Gender: Women's Changing Participation in Computing*. Cambridge, Mass: MIT Press.

Almeida, Virgilio, Danilo Doneda, and Carolina Rossini. 2016. How Do App Stores Challenge the Global Internet Governance Ecosystem? *IEEE Internet Computing* 20 (6): 49–51. https://doi.org/10.1109/MIC.2016.123.

Alston, Philip. 2017. The Populist Challenge to Human Rights. *Journal of Human Rights Practice* 9 (1): 1–15. https://doi.org/10.1093/jhuman/hux007.

Antonova, Slavka. 2013. Internet and the Emerging Global Community of Rights: The Human Rights Debate at the Internet Governance Forum. *Journal of Philosophy of International Law* 4 (01): 84.

Appelman, Daniel L. 2016. Internet Governance and Human Rights: ICANN's Transition Away from United States Control. The Clarion, a Journal of the American Bar Association's International Human Rights Committee, 1(1).

Badii, Farzaneh, and Milton Mueller. 2018. Requiem for a Dream: On Advancing Human Rights via Internet Architecture. *Policy & Internet*.

Balkin, Jack M. forthcoming. Free Speech in the Algorithmic Society: Big Data, Private Governance, and New School Speech Regulation. *UC Davies Law Review*. https://papers.ssrn.com/abstract=3038939.

Barlow, John Perry. 1996. *A Declaration of the Independence of Cyberspace*. https://projects.eff.org/~barlow/Declaration-Final.html.

Baumann-Pauly, Dorothée, Justine Nolan, Auret Heerden, and Michael Samway. 2017. Industry-Specific Multi-Stakeholder Initiatives That Govern Corporate Human Rights Standards: Legitimacy Assessments of the Fair Labor Association and the Global Network Initiative. *Journal of Business Ethics* 143 (4): 771–787. https://doi.org/10.1007/s10551-016-3076-z.

Benedek, Wolfgang. 2008. Internet Governance and Human Rights. In *Internet Governance and Information Society*, ed. Wolfgang Benedek, Veronika Bauer, and Matthias Kettemann. Utrecht: Eleven International Publishing.

Benkler, Yochai. 2000. From Consumers to Users: Shifting the Deeper Structures of Regulation Towards Sustainable Commons and User Access. *Federal Communications Law Journal* 52 (3). https://dash.harvard.edu/handle/1/11363059.

———. 2006. *The Wealth of Networks: How Social Production Transforms Markets and Freedom*. New Haven and London: Yale University Press.

Belli, Luca. 2017. *Network self-determination and the positive externalities of community networks*. http://bibliotecadigital.fgv.br/dspace/handle/10438/19924.

Berners-Lee, Tim. 2014. *Recognise the Internet as a human right, says Sir Tim Berners-Lee as he launches annual Web Index*. Retrieved from https://webfoundation.org/2014/12/recognise-the-internet-as-a-human-right-says-sir-tim-berners-lee-as-he-launches-annual-web-index/.

boyd, danah. 2012. White Flight in Networked Publics? How Race and Class Shaped American Teen Engagement with MySpace and Facebook. In *In Race After the Internet*, ed. Lisa Nakamura and Peter Chow-White. London: Routledge.

Braman, Sandra. 2011. Privacy by Design: Networked Computing, 1969–1979. *New Media & Society* 14 (5): 798–814. https://doi.org/10.1177/1461444811426741.

———. 2017. Internet Histories: The View from the Design Process. *Internet Histories* 1 (1–2): 70–78. https://doi.org/10.1080/24701475.2017.1305716.

Brousseau, Eric, Meryam Marzouki, and Cecile Meadel. 2012. *Governance, regulation and powers on the internet - Cambridge University press*. Cambridge, UK: Cambridge University Press. http://www.cambridge.org/catalogue/catalogue.asp?isbn=9781107502611&ss=cop.

Brown, Ian, and Christopher T. Marsden. 2013. *Regulating Code: Good Governance and Better Regulation in the Information Age*. Cambridge, MA: MIT Press.

Brown, Ian, David Clark, and Dirk Trossen. 2010. *Should Specific Values Be Embedded In The Internet Architecture?* Vol. Paper presented at th Re-Architecting the Internet workshop. New York. http://conferences.sigcomm.org/co-next/2010/Workshops/REARCH/ReArch_papers/10-Brown.pdf.

Burrington, Ingrid. 2016. *Networks of New York: An Internet Infrastructure Field Guide*. http://seeingnetworks.in/guide/.

Bygrave, Lee A. 2012. Contract Versus Statute in Internet Governance. In *Research Handbook on Governance of the Internet, edited by Ian Brown*. Cheltenham: Edward Elgar. https://papers.ssrn.com/abstract=1948518.

———. 2017. *Hardwiring Privacy*, SSRN Scholarly Paper ID 2901405. Rochester: Social Science Research Network. https://papers.ssrn.com/abstract=2901405.

Bygrave, Lee A., and Jon Bing, eds. 2009. *Internet Governance: Infrastructure and Institutions*. 1st ed. Oxford/New York: Oxford University Press.

Carr, Madeline. 2013. Internet Freedom, Human Rights and Power. *Australian Journal of International Affairs*, 1–17. https://doi.org/10.1080/10357718.2013.817525.

———. 2015. Power Plays in Global Internet Governance. *Millennium – Journal of International Studies* 43 (2): 640–659. https://doi.org/10.1177/0305829814562655.

Cath, C. forthcoming. The Technology we choose to create: Human rights advocacy and anthropology in internet governance. Conference Proceedings of GigaNet International Communications Association Preconference 2019.

Cath, Corinne, and Luciano Floridi. 2017. The Design of the Internet's architecture by the internet engineering task force (IETF) and human rights. *Science and Engineering Ethics* 23 (2): 449–468.

Castells, Manuel. 2000. *The Rise of The Network Society: The Information Age: Economy*. Society and Culture: Wiley.

Cavoukian, Ann. 2012. Privacy by Design [Leading Edge]. *IEEE Technology and Society Magazine* 31 (4): 18–19. https://doi.org/10.1109/MTS.2012.2225459.

Cerf, V.G. 2012. Emergent properties, human rights, and the internet. *Internet Computing, IEEE* 16 (2): 87–88. https://doi.org/10.1109/MIC.2012.32.

Choucri, Nazli, and David D. Clark. 2012. Who Controls Cyberspace? *Bulletin of the Atomic Scientists* 69 (5): 21–31. https://doi.org/10.1177/0096340213501370.

Christou, George, and Seamus Simpson. 2009. New Governance, the Internet, and Country Code Top-Level Domains in Europe. *Governance* 22 (4): 599–624. https://doi.org/10.1111/j.1468-0491.2009.01455.x.

Cong, Wanshu. 2017. Understanding Human Rights on the Internet: An Exercise of Translation? *Tilburg Law Review* 22 (1–2): 138–164. https://doi.org/10.1163/22112596-02201007.

Deibert, Ronald. 2015. The Geopolitics of Cyberspace After Snowden. *Current History*. https://www.researchgate.net/publication/279327884_The_Geopolitics_of_Cyberspace_After_Snowden.

Deibert, Ronald, John Palfrey, Rafal Rohonzinski, Jonathan Zittrain, and Janice Stein. 2008. *Access Denied: The Practice and Policy of Global Internet Filtering*. Boston: MIT Press.

DeNardis, Laura. 2012. Hidden Levers of Internet Control. *Information, Communication & Society* 15 (5): 720–738. https://doi.org/10.1080/1369118X.2012.659199.

———. 2013. *Protocol Politics: The Globalization of Internet Governance*. Boston: MIT Press.

———. 2014. *The Global War for Internet Governance*. New Haven: Yale University Press.

DeNardis, Laura, and Andrea Hackl. 2016. Internet Control Points as LGBT Rights Mediation. *Information, Communication & Society* 19 (6): 753–770. https://doi.org/10.1080/1369118X.2016.1153123.

Dourish, Paul. 2015. Protocols, Packets, and Proximity: The Materiality of Internet Routing. In *Signal Traffic: Critical Studies of Media Infrastructure, edited by Lisa Parks and Nicole Starosielski*. Illinois: University of Illinois Press.

Drake, William, and Ernest Wilson, eds. 2004. *Governing Global Electronic Networks: International Perspectives on Policy and Power*. Boston: MIT Press. https://mitpress.mit.edu/books/governing-global-electronic-networks.

Drezner, Daniel. 2004. The Global Governance of the Internet: Bringing the State Back In. *Political Science Quarterly* 119: 447–498.

Dutton, William, and Malcolm Peltu. 2005. The Emerging Internet Governance Mosaic: Connecting the Pieces. *Forum Discussion Paper No. 5 Oxford Internet Institute*.

Easton, Catherine. 2016. *Internet Governance: A Human Rights Perspective*. 1st ed. London: Routledge.

Epstein, Dmitry, Christian Katzenbach, and Francesca Musiani. 2016. Doing Internet Governance: Practices, Controversies, Infrastructures, and Institutions. *Internet Policy Review*. https://policyreview.info/articles/analysis/doing-internet-governance-practices-controversies-infrastructures-and-0.

Eubanks, Virginia. 2018. *Automating Inequality: How High-Tech Tools Profile, Police, and Punish the Poor*. New York: St. Martin's Press.

Floridi, Luciano. 2014. *The Fourth Revolution: How the Info sphere Is Reshaping Human Reality*. Oxford/New York: Oxford University Press.

———. 2017. Digital's Cleaving Power and Its Consequences. *Philosophy & Technology* 30 (2): 123–129. https://doi.org/10.1007/s13347-017-0259-1.

Flyverbom, Mikkel. 2010. Hybrid Networks and the Global Politics of the Digital Revolution – A Practice-Oriented, Relational and Agnostic Approach. *Global Networks* 10 (3): 424–442. https://doi.org/10.1111/j.1471-0374.2010.00296.x.

Friedman, Batya, Peter H. Kahn, Alan Borning, and Alina Huldtgren. 2013. Value Sensitive Design and Information Systems. In *Early Engagement and New Technologies: Opening up the Laboratory*, 55–95. Dordrecht: Philosophy of Engineering and Technology. Springer. https://doi.org/10.1007/978-94-007-7844-3_4.

Gill, Lex, Dennis Redeker, and Urs Gasser. 2015. Towards Digital Constitutionalism? Mapping Attempts to Craft an Internet Bill of Rights. *Berkman Center Research Publication*. https://papers.ssrn.com/abstract=2687120.

Gillespie, Tarleton. 2010. The Politics of "Platforms". *New Media & Society* 12 (3): 347–364. https://doi.org/10.1177/1461444809342738.

Goldsmith, Jack. 1998. The Internet and the Abiding Significance of Territorial Sovereignty. *Indiana Journal of Global Legal Studies* 5 (2).

Goldsmith, Jack, and Wu. Tim. 2006. *Who Controls the Internet? Illusions of a Borderless World*. Oxford: Oxford University Press. http://jost.syr.edu/wp-content/uploads/who-controls-the-internet_illusions-of-a-borderless-world.pdf.

Graham, Mark, Isis Hjorth, and Vili Lehdonvirta. 2017. Digital Labour and Development: Impacts of Global Digital Labour Platforms and the Gig Economy on Worker Livelihoods. *Transfer: European Review of Labour and Research* 23 (2): 135–162. https://doi.org/10.1177/1024258916687250.

Hackl, Andrea. 2016. Internet Policy Designs as "Infrastructures of LGBTQ Expression"- Internet Governance as a Minority Rights Issue. Pro Quest Dissertations Publishing. http://search.proquest.com/docview/1865331070/.

Helmond, Anne. 2015. The Platformization of the Web: Making Web Data Platform Ready. *Social Media + Society* 1 (2): 2056305115603080. https://doi.org/10.1177/2056305115603080.

Hick, Steven, Edward Halpin, and Eric Hoskins, eds. 2000. *Human Rights and the Internet*. Suffolk: Palgrave Schol.

Hicks, Marie. 2017. *Programmed Inequality: How Britain Discarded Women Technologists and Lost Its Edge in Computing*. Cambridge, MA/London: MIT Press.

Hintz, Arne, and Stefania Milan. 2009. At the Margins of Internet Governance: Grassroots Tech Groups and Communication Policy. *International Journal of Media & Cultural Politics* 5: 23–38.

Hofmann, Jeanette. 2007. *Internet Governance: A Regulative Idea in Flux*. Edited by Ravi Kumar Jain Bandamutha. In Internet Governance: An Introduction. Rochester. https://papers.ssrn.com/abstract=2327121.

Hofmann, Jeanette, Christian Katzenbach, and Kirsten Gollatz. 2017. Between Coordination and Regulation: Finding the Governance in Internet Governance. *New Media & Society* 19 (9): 1406–1423. https://doi.org/10.1177/1461444816639975.

Johnson, David, and David Post. 1997. And How Shall the Net Be Governed? A Mediation on the Relative Virtues of Decentralized, Emergent Law. In *In Coordinating the Internet*, ed. Brian Kahin and James Keller. Cambridge: MIT Press.

Jørgensen, Rikke Frank, ed. 2006. *Human Rights in the Global Information Society*. Boston: MIT Press. https://mitpress.mit.edu/books/human-rights-global-information-society.

———. 2013. *Framing the Net*. Cheltenham: Edward Elgar Publishing.

———. 2017. What Platforms Mean When They Talk About Human Rights. *Policy & Internet* 9 (3): 280–296. https://doi.org/10.1002/poi3.152.

———, ed. forthcoming. *Human Rights in the Age of Platforms*. Boston: MIT Press.

Jørgensen, Rikke, and Meryem Marzouki. 2015. *Internet Governance and the Reshaping of Global Human Rights Legacy at WSIS + 10*, SSRN Scholarly Paper ID 2809888. Rochester: Social Science Research Network. https://papers.ssrn.com/abstract=2809888.

Joyce, Daniel. 2015. Internet Freedom and Human Rights. *European Journal of International Law* 26 (2): 493–514. https://doi.org/10.1093/ejil/chv021.

Kahin, Brian, and James H. Keller. 1997. The Problem of Internet Governance. In *Coordinating the Internet*, 1–1. MIT Press. http://ieeexplore.ieee.org/xpl/articleDetails.jsp?arnumber=6275239.

Karanicolas, Michael. 2013. Bridging the Divide: Understanding and Implementing Access to the Internet as a Human Right. *The Journal of Community Informatics* 10 (2). http://ci-journal.net/index.php/ciej/article/view/990.

Kiggins, David. 2015. Open for Expansion: US Policy and the Purpose for the Internet in the Post–Cold War Era. *International Studies Perspectives* 16 (1). http://onlinelibrary.wiley.com/wol1/doi/10.1111/insp.12032/full.

Klang, Mathias, and Andrew Murray. 2004. *Human Rights in the Digital Age*. 1st ed. London/Portland: Routledge-Cavendish.

Klonick, Kate. forthcoming. The New Governors: The People, Rules, and Processes Governing Online Speech. *Harvard International Law Journal* 131. https://papers.ssrn.com/abstract=2937985.

Kopetz, Hermann. 2011. Internet of Things. In *Real-Time Systems*, 307–323. Boston: Real-Time Systems Series. Springer. https://doi.org/10.1007/978-1-4419-8237-7_13.

Kulesza, Joanna. 2008. Freedom of Information in the Global Information Society: The Question of the Internet Bill of Rights. *Law Review* 1. https://papers.ssrn.com/abstract=1446771.

Laidlaw, Emily. 2012. *Internet Gatekeepers, Human Rights and Corporate Social Responsibilities*. http://etheses.lse.ac.uk/317/1/Laidlaw_Internet%20Gatekeepers%2C%20Human%20Rights%20and%20Corporate%20Social%20Responsibilities.pdf.

Land, Molly K. 2013. Toward an International Law of the Internet. *Harvard International Law Journal* 54. https://papers.ssrn.com/abstract=2177993.

Land, Molly K., and Jay Aronson, eds. 2018. *The Challenges and Opportunities of New Technologies*. Cambridge: Cambridge University Press.

Lessig, Lawrence. 2004. *Free Culture: How Big Media Uses Technology and the Law to Lock Down Culture and Control Creativity*. London: Penguin Books.

———. 2006. *Code: And Other Laws of Cyberspace, Version 2.0*. New York: Basic Books.

Lettinga, Doutje, and Lars Van Troost, eds. 2015. Can Human Rights Bring Social Justice? 12 Essays. *Amnesty International Netherlands*. http://www.academia.edu/16396080/Can_human_rights_bring_social_justice_12_essays.

Lucchi, Nicola. 2014. Internet Content Governance and Human Rights. *Vanderbilt Journal of Entertainment and Technology Law* 16 (4): 809–856.

MacKinnon, Rebecca. 2012. *Consent of the Networked: The Worldwide Struggle For Internet Freedom*. New York: Basic Books.

Maréchal, Nathalie. 2015. Ranking Digital Rights: Human Rights, the Internet and the Fifth Estate. *International Journal of Communication* 9 (10): 3440–3449.

———. 2017. Networked Authoritarianism and the Geopolitics of Information: Understanding Russian Internet Policy. *Media and Communication* 5 (1): 29–41. https://doi.org/10.17645/mac.v5i1.808.

Marsden, Chris. 2008. Beyond Europe: The Internet, Regulation, and Multistakeholder Governance--Representing the Consumer Interest? *Journal of Consumer Policy* 31 (1). https://ezproxy-prd.bodleian.ox.ac.uk:7316/ibss/docview/198366277/BC92085A22F64407PQ/185?accountid=13042.

Mathew, Ashwin J. 2016. The Myth of the Decentralised Internet. *Internet Policy Review*. https://policyreview.info/articles/analysis/myth-decentralised-internet.

Mathiason, John. 2008. *Internet Governance: The New Frontier of Global Institutions*. London: Routledge.

Mathiesen, Kay. 2014. Human Rights for the Digital Age. *Journal of Mass Media Ethics* 29 (1): 2–18. https://doi.org/10.1080/08900523.2014.863124.

McGoldrick, Dominic. 2013. Developments in the Right to Be Forgotten. *Human Rights Law Review* 13 (4): 761–776. https://doi.org/10.1093/hrlr/ngt035.

McIlwain, Charlton. 2016. Racial Formation, Inequality and the Political Economy of Web Traffic. *Information, Communication & Society* 20 (7). https://www.tandfonline.com/doi/abs/10.1080/1369118X.2016.1206137.

Meier-Hahn, Uta. 2015. Creating Connectivity: Trust, Distrust and Social Microstructures at the Core of the Internet. *HIIG Discussion Paper Series* No. 2015–03. https://papers.ssrn.com/abstract=2587843.

Metzl, Jamie F. 1996. Information Technology and Human Rights. *Human Rights Quarterly* 18 (4): 705–746.

Milan, Stefania, and Niels Ten Oever. 2016. Coding and Encoding Rights in Internet Infrastructure. *Internet Policy Review* 6 (1). https://doi.org/10.14763/2017.1.442.

Moller, Christian. 2013. New Technology, Minorities and Internet Governance. *Journal on Ethnopolitics and Minority Issues in Europe* 12: 16–33.

Moyn, Samuel. 2012. *The Last Utopia*. Boston: Harvard University Press. http://www.hup.harvard.edu/catalog.php?isbn=9780674064348.

Mueller, Milton. 2002. *Ruling the Root: Internet Governance and the Taming of Cyberspace*. Cambridge MA: MIT Press.

Mueller, Milton. 2016. Missing the target: The human rights push in ICANN Goes off the rails. *Internet Governance Project (blog)*. 2016. https://www.internetgovernance.org/2016/10/26/missing-the-target-the-humanrights-push-in-icann-goes-off-the-rails/.

———. 2010. *Networks and States*. Cambridge MA: MIT Press.

Mueller, Milton, and Ralf Bendrath. 2010. 'The End of the Net as We Know It? Deep Packet Inspection and Internet Governance by Ralf Bendrath, Milton Mueller :: SSRN'. *The Oxford Handbook on Internet Studies*. https://papers.ssrn.com/sol3/papers.cfm?abstract_id=1653259.

Mueller, Milton, and Farzaneh Badiei. 2018. Requiem for a dream: On advancing human rights via internet architecture. *Policy & Internet* (0). https://doi.org/10.1002/poi3.190.

Mueller, Milton, John Mathiason, and Hans Klein. 2007. The Internet and Global Governance: Principles and Norms for a New Regime. *Global Governance* 13 (2): 237–254.

Musiani, Francesca. 2012. A Bill of Rights for Internet Users? *Droit et Societe* 81. https://ezproxy-prd.bodleian.ox.ac.uk:7316/ibss/docview/1418130433/BC92085A22F64407PQ/270?accountid=13042.

Musiani, Francesca, Derrick L. Cogburn, Laura DeNardis, and Nanette S. Levinson. 2015. *The Turn to Infrastructure in Internet Governance*. New York: Palgrave Macmillan US.

Naughton, John. 1999. *A Brief History of the Future: The Origins of the Internet*. London: Weidenfeld & Nicolson.

Nielsen, Rasmus Kleis, and Sarah Anne Ganter. 2017. Dealing with Digital Intermediaries: A Case Study of the Relations between Publishers and Platforms. *New Media & Society*. https://doi.org/10.1177/1461444817701318.

Noble, Safiya Umoja. 2018. *Algorithms of Oppression: How Search Engines Reinforce Racism*. New York: NYU Press.

Orwat, Carsten, and Roland Bless. 2016. Values and Networks: Steps Toward Exploring Their Relationships. *ACM SIGCOMM Computer Communication Review* 46 (2): 25–31.

Padovani, Claudia, Francesca Musiani, and Elena Pavan. 2010. Investigating Evolving Discourses On Human Rights in the Digital Age. *International Communication Gazette* 72 (4–5): 359–378. https://doi.org/10.1177/1748048510362618.

Pettrachin, Andrea. 2018. Towards a Universal Declaration on Internet Rights and Freedoms? *International Communication Gazette*. https://doi.org/10.1177/1748048518757139.

Powers, Shawn, and Michael Jablonski. 2015. *The Real Cyber War: The Political Economy of Internet Freedom*. Urbana: University of Illinois Press. http://www.jstor.org/stable/10.5406/j.ctt130jtjf.

Rachovitsa, Adamantia. 2016. Rethinking Privacy Online and Human Rights: The Internet's Standardisation Bodies as the Guardians of Privacy Online in the Face of Mass Surveillance. *European Society of International Law*.

Raymond, Mark, and Laura DeNardis. 2015. Multistakeholderism: Anatomy of an Inchoate Global Institution. *International Theory* 7 (3): 572–616. https://doi.org/10.1017/S1752971915000081.

Redeker, Dennis, Lex Gill, and Urs Gasser. 2018. Towards Digital Constitutionalism? Mapping Attempts to Craft an Internet Bill of Rights. *International Communication Gazette*, 1748048518757121. doi:https://doi.org/10.1177/1748048518757121.

Reidenberg, Joel. 1997. Lex Informatica: The Formulation of Information Policy Rules through Technology. *Texas Law Review* 76 (3): 553–593.

Ross, Margaret. 2014. Understanding Interconnectivity of the Global Undersea Cable Communications Infrastructure and Its Implications for International Cyber Security. *The SAIS Review of International Affairs* 34 (1). https://ezproxy-prd.bodleian.ox.ac.uk:7316/ibss/docview/1552151690/BC92085A22F64407PQ/4?accountid=13042.

Ruggie, John. 2017. Multinationals as Global Institution: Power, Authority and Relative Autonomy. *Regulation & Governance*. https://doi.org/10.1111/rego.12154.

Rundle, Mary, and Malcolm Birdling. 2008. Filtering and the International System. In *In Access Denied: The Practice and Policy of Global Internet Filtering*, ed. Ronald Deibert, John Palfrey, Rafal Rohonzinski, and Jonathan Zittrain. Boston: MIT Press.

Santoro, Mauricio, and Bruno Borges. 2017. Brazilian Foreign Policy Towards Internet Governance. *Revista Brasileira de Politica Internacional* 60 (1). https://ezproxy-prd.bodleian.ox.ac.uk:7316/ibss/docview/1867921464/BC92085A22F64407PQ/129?accountid=13042.

Scholte, Jan Aarte. forthcoming. Complex Hegemony: The IANA Transition in Global Internet Governance. *Giga Net Symposium IGF Geneva 2017*.

Schuster, Stefan, Melle Van Den Berg, Xabier Larrucea, Ton Slewe, and Peter Ide-Kostic. 2017. Mass Surveillance and Technological Policy Options: Improving Security of Private Communications. *Computer Standards & Interfaces* 50: 76–82.

Shackelford, Scott J., Enrique Oti, Jaclyn A. Kerr, Elaine Korzak, and Andreas Kuehn. 2015. Back to the Future of Internet Governance Global Governance. *Georgetown Journal of International Affairs* 16: 83–97.

Smith, Brad. 2017. *The Need for a Digital Geneva Convention*. Retrieved from https://blogs.microsoft.com/on-the-issues/2017/02/14/need-digital-genevaconvention/.

Solum, Lawrence B. 2009. Models of Internet Governance'. In *Internet Governance: Infrastructures and Institutions*, edited by Lee A. Bygrave and Jon Bing. https://papers.ssrn.com/abstract=1136825.

Starosielski, Nicole. 2015. *The Undersea Network*. Durham: Duke University Press Books.

Suzor, Nicolas, Tess Van Geelen, and Sarah Myers West. 2018. Evaluating the Legitimacy of Platform Governance: A Review of Research and a Shared Research Agenda. *International Communication Gazette*, 1748048518757142. https://doi.org/10.1177/1748048518757142.

Taylor, Linnet, Luciano Floridi, and Bart van der Sloot. 2017. *Group Privacy - New Challenges of Data Technologies*. Berlin: Springer. www.springer.com/gb/book/9783319466064.

Torraco, Richard J. 2005. Writing Integrative Literature Reviews: Guidelines and Examples. *Human Resource Development Review* 4 (3): 356–367. https://doi.org/10.1177/1534484305278283.

———. 2016. Writing Integrative Literature Reviews. *Human Resource Development Review* 15 (4): 404.

Tully, Stephen. 2014. A Human Right to Access the Internet? Problems and Prospects. *Human Rights Law Review* 14 (2): 175–195. https://doi.org/10.1093/hrlr/ngu011.

van Eeten, Michel J.G., and Milton Mueller. 2013. Where Is the Governance in Internet Governance? *New Media & Society* 15 (5): 720–736. https://doi.org/10.1177/1461444812462850.

Van Hoboken, Joris. 2012. *Search Engine Freedom. On the Implications of the Right to Freedom of Expression for the Legal Governance of Web Search Engines*. Alphen aan de Rijn: Kluwer Law International.

Vezzani, Simone. 2014. ICANN's New Generic Top-Level Domain Names Dispute Resolution Procedure Viewed against the Protection of the Public Interest of the Internet Community: Litigation Regarding Health-Related Strings. *Law and Practice of International Courts and Tribunals* 13 (3): 306–346. https://doi.org/10.1163/15718034-12341279.

Wiener, Norbert. 1954. *The Human Use of Human Beings Cybernetics and Society*. Garden City: Doubleday Anchor Books. http://dspace.nehu.ac.in/handle/123456789/12979.

Williams, James. forthcoming. *Stand Out of Our Light: Freedom and Persuasion in the Attention Economy*. Cambridge: Cambridge University Press.

Ziewitz, Malte, and Christian Pentzold. 2014. In Search of Internet Governance: Performing Order in Digitally Networked Environments. *New Media & Society* 16 (2): 306–322. https://doi.org/10.1177/1461444813480118.

Zittrain, Jonathan. 2008. *The Future of the Internet - And How to Stop It*. New Haven: Yale University Press.

Chapter 9
Privacy Risks and Responses in the Digital Age

Josh Cowls

Digital devices, the data they collect, and the algorithms that process this data have transformed government, business and everyday life. In this chapter I offer a brief introduction to privacy in the context of American jurisprudence. I then identify major risks to personal privacy that occur at three stages of the "lifecycle" of data: infrastructural asymmetry at the point of data collection, distortion at the point of data analysis, and discrimination at the point of deployment of the insights of data analysis. Having identified these risks, I then introduce a framework consisting of four categories of responses that can be adopted to mitigate them. These categories of responses apply at the level of the individual and of society at large, and in both instrumental and epistemological senses. I conclude by arguing that this framework should serve as a basis for future research into the effects of ICTs on privacy.

9.1 Introduction

The mass adoption of digitally-connected devices has created new commercial, political and social significance for the avalanche of data they collect and the insights they yield. Recent high-profile legal disputes between, for example, intelligence agencies and technology companies (Nakashima 2016), or individuals and search engines (Lomas 2017), hint at broader tussles between different sectors of society over the control of personal data. At the heart of these skirmishes lies the principle of informational privacy (hereafter just privacy), the threats to which have been exacerbated by digital information and communication technologies (Floridi 2005).

J. Cowls (✉)
Oxford Internet Institute, Digital Ethics Lab, University of Oxford, Oxford, UK

The Alan Turing Institute, London, UK
e-mail: josh.cowls@oii.ox.ac.uk

© Springer Nature Switzerland AG 2019
C. Öhman, D. Watson (eds.), *The 2018 Yearbook of the Digital Ethics Lab*,
Digital Ethics Lab Yearbook, https://doi.org/10.1007/978-3-030-17152-0_9

The heightened threats to personal privacy introduced by the modern devices and the data that they capture are, by themselves, no longer especially new. Much of the infrastructure undergirding CCTV surveillance, for example, has been in place in many countries for decades, while this generation's most popular smartphone, the iPhone, was first introduced in 2007. More recent, and potentially riskier, however, are the new opportunities for analysis afforded by automated decision-making systems. Such systems are fuelled by increasingly abundant data, and their sophistication is vastly increased through the leveraging of complex machine learning techniques such as deep neural network approaches. These new algorithmic systems can be deployed on data collected by devices many years earlier.

As such, it is the aggregation of (i) ubiquitous digital devices, (ii) the data they collect, and, most significantly (iii) the decision-making that algorithmic systems enable which I address in this paper. Each of these technological developments has generated new and qualitatively distinct threats to privacy, and each successive development has served to compound and augment earlier effects.

I begin by sketching a brief background to the right to privacy, with particular reference to American jurisprudence, arguing that evolving conceptions of privacy have long been interwoven with technological developments. I then introduce three core threats to privacy that emerge at each stage of the data "lifecycle". Finally, I offer a series of possible responses to these threats, organised in a two-dimensional framework offering sets of instrumental and epistemological approaches at micro and macro levels. I conclude by arguing that this framework should serve as a basis for future research into the effects of ICTs on privacy.

9.2 Privacy in an Age of Technology

9.2.1 The Emergence of a Right to Privacy

In a classic article of 1890, Warren and Brandeis espoused "the right to privacy" (1890), probably for the first time in the history of American jurisprudence (Glancy 1979). Warren and Brandeis argued that "recent inventions and business methods call attention to the [need] for securing to the individual … the right 'to be let alone'" (p. 195). Whereas in 1890, the chief threat to personal privacy was posed by "instantaneous photographs [used by] newspaper enterprises" (p. 195), the 1928 Supreme Court case *Olmstead v. United States*, which Brandeis heard as an associate justice, turned instead on whether government agents had violated the constitution by securing evidence through the use of a wire-tap of a telephone call.

It is clear, even from this early judicial skirmish,[1] that privacy as a concept is thoroughly bound up in the control of the flow of sensitive, meaningful or in other

---

[1] In the interest of brevity, I restrict my discussion of the history of privacy to the American context, though there are of course antecedents in other countries. For example, support for a right to privacy is found in Article 8 of the European Convention on Human Rights, and in India following Puttaswamy vs Union of India (2017; see also Panday 2017).

ways valuable information—and the intersection of this information with personal identity. In their first expression of a right to privacy, Warren and Brandeis offer the example of a private diary in which a man records the fact that he did not on a given evening dine with his wife. This simple but potentially inconvenient truth has no special artistic merit or value, and so cannot be protected as a form of (intangible) property, as would a patent or musical composition. What is protected "is not the intellectual act of recording the fact that the husband did not dine with his wife, but that fact itself" (1890, p. 201; italics added). Much more recently, Floridi (2016) echoes this argument, suggesting that the protection of privacy is *ipso facto* the protection of personal identity, since personal information plays a constitutive role in an individual's identity. The evolution of the mooted right to privacy, from its first emergence in Brandeis's jurisprudence in 1890 to its application in American law in 1967, has thus developed, at least in part, as a reflection of the changing technology available to both the citizenry and the state. In his original 1928 dissent, Brandeis argued that there should be no judicial distinction between a sealed letter—which already enjoyed constitutional protection—and a telephone conversation, supporting the idea that constitutional protections should keep pace with technological development.

In *Olmstead*, the Court ultimate ruled 5–4 that the alleged violation had not occurred. Yet Brandeis's dissenting opinion was a striking elucidation of the nascent "right to privacy", setting the stage for more than a century of evolving understanding of how a right to privacy could be justified and implemented. Brandeis's dissent was ultimately vindicated in *Katz v. United States* (1967), which overturned *Olmstead* by asserting that Fourth Amendment protection against unreasonable search and seizure extended to *electronic* intrusion. As in *Olmstead*, the *Katz* case turned on the validity of evidence obtained by wire-tapping conducted by federal agents. In *Katz*, even though the suspect had conducted incriminatory conversations on a pay phone in a public space, the Court ruled that he had a "reasonable expectation of privacy" by taking such simple measures as closing the door of the phone booth behind him. In this enclosed space, the suspect was "entitled to assume that the words he utters into the mouthpiece will not be broadcast to the world" (p. 352).

Even if we can find in Brandeis's jurisprudence the genesis of a right to privacy (at least in United States law), today we are still a long way from an agreed-upon understanding of how it might be defined. As Judith Jarvis Thomson notes, "the most striking thing about the right to privacy is that nobody seems to have any very clear idea what it is" (Thomson 1975, p. 295), even though, as Richard Posner has remarked, "much ink has been spilled in trying to clarify its meaning" (1978). Privacy has elsewhere been characterised as an umbrella term (Solove 2005, p. 485). Privacy, then, either means a lot of different things, or not much at all—neither a good basis for an agreed-upon definition. That most privacy law was developed before the most recent wave of technological innovation only complicates the picture.

Nonetheless, attempts have been made to impose order on the amorphous principle of privacy. Daniel Solove's authoritative "Taxonomy of Privacy" (2005) is structured around the activities that invade privacy: information collection; information processing; information dissemination; and invasion. Solove's focus on *activities* as the basis of privacy harms adapts well to the digital age, and particularly to my own framing. Solove argues, for instance, that there are cases in which the mere collection of information about a person can cause a privacy harm, as in the case of surveillance—an argument more readily apparent in an age of the bulk collection of communications alleged by Edward Snowden's disclosures in 2013. The contextual considerations around privacy as it relates to the digital age have also been notably emphasised by Nissenbaum (2009).

Regardless of how it is conceptualised, the value of privacy has tended to be weighed against, or in relation to, other considerations. Brandeis himself recognised exceptions to privacy, particularly as it related to the public interest. In their 1890 article, Warren and Brandeis argued that matters which should remain private are those which "have no legitimate connection with [an individual's] fitness for a public office" (p. 216). And in *Other People's Money And How the Bankers Use It*, Brandeis noted that "sunlight is said to be the best of disinfectants; electric light the most efficient policeman" against the consolidated power of the oligarchic "Money Trust"—adding further weight to the notion that the right to individual privacy and the utility of transparency can and should coexist (2009[1914], p. 62).

9.2.2   Privacy and Modern Technology

Successive waves of post-war innovation have fundamentally altered the guarantee of individual privacy that Brandeis sought to inscribe. From the rollout of CCTV cameras and the use of wiretapping, to the more recent avalanche of personal data generated by widespread use of connected devices and platforms today, threats to the privacy of individuals in modern society are without historical precedent. Indeed, as Floridi (2017, p. 123) has suggested, it is "the digital, with its extended power to record, monitor, share, and process endless amounts of data" that has "soldered together who [someone] is" with their "personal information".

Things are becoming more complicated still with the rise of artificial intelligence, which empowers non-human agents to make data-driven decisions. This has created an aggregation effect, whereby the effects of more recent technologies are augmented or layered atop existing technological infrastructure to create new privacy risks. An example may serve to elucidate this. It has recently been revealed that artificial intelligence technologies developed by Google are being used to sort through the masses of video footage shot by US military drones (Gibbs 2018). The controversy surrounding Google's involvement in this activity rests on the power of its AI technology to augment the scores of expensive, sophisticated drones operated by the US Department of Defense, and the vast stores of high-quality video data that they are collecting. Previously, the task of analysing all this video had rested with humans.

Where, in this example, might the harm to personal privacy be located? There is no obvious single answer. Some people may quite reasonably object in principle to a swarm of unmanned aircraft surveilling their home or workplace. For others, it might be the fact that these aircraft are collecting many hours of video, perhaps capturing people's movement on a day to day level. For others still, it might be the fact that—as it now transpires—all this video footage is being pored over and probed for meaning by sophisticated algorithms at super-human speeds. The issue (at least, within the scope of this paper) is less where in particular the harm should be said to occur, and more that each successive wave of technological development— devices, data, and automated decision-making systems—builds upon and compounds the effects of what came before.

Moreover, while these technologies are logically ordinal—at least insofar as data capture relies on devices, and automated decision-making is powered by available data—in practical terms threats to privacy can arise from quite diverse timescales. It may be that a sufficiently suspicious tweet posted tonight would earn a visit from security services before sunrise. But it is also conceivable that the next generation of political leaders—those currently in their teens and twenties—will face considerable scrutiny into questionable behaviour that occurred many years earlier. A scandalous photo taken and published in 2010 on a device purchased in 2008 may not emerge until a suitably efficient facial recognition algorithm is developed in 2025.

This example serves to suggest that the power to control the flow of information is neither gained nor exercised in a vacuum. It rests on a complex set of legal, political, and economic forces in society. Often these powers are explicitly defined, as with data protection and freedom of information legislation, which rest on the principles of privacy and transparency, respectively. Yet a raft of more oblique but no less significant factors also have an impact, from the ability of individuals to understand a privacy policy, to the impact of a biased algorithm on a loan decision. This suggests, anticipating my conclusion, that a range of responses, from "hard" laws to "softer" solutions such as improved public understanding, are required to counter these multifarious threats.

9.3   Privacy Risks Across the "Lifecycle" of Data

In the previous section I argued that the emerging rights to privacy have long been bound up with evolving technologies, from the end of the nineteenth century to the present. In this section I introduce three sets of risks to privacy that emerge across the "lifecycle" of data in contemporary life. Privacy risks concerning the use and misuse of digital devices, data, and decision-making systems are best captured by three distinct provocations. The first risk is the asymmetry in information infrastructure as it relates to data *collection*; in other words, *who* has the infrastructural ability to capture and use data? Whether the state observes its citizens, or whether citizens are able to observe their state, are outcomes of the power relations in a given society. There are many other configurations too, such as corporations selling advertisements based on an intimate knowledge of an individual's interests, as harvested from their data. The second risk I introduce here is the *distortion* that can occur at the point of data analysis; in other words, *how* is data utilised, and with what effect? This consideration is vital because much of the value of data rests on the insights that can be drawn from it. The third risk introduced is discrimination in the differential deployment of data-driven insights across particular populations; in other words, *how and where* are the insights of data analysis deployed, and with what implications for privacy?

This framework of risks is by no means exhaustive; it is, however, both comprehensive and concise. It covers the entire lifecycle of data processing, from capture and retention to analysis and finally the deployment of insights.[2] In the remainder of this section I explore these risks in turn.

9.3.1 Infrastructural Asymmetry at the Point of Data Collection

The notion of "transparency" has become inseparable and almost synonymous with the rise of what has been dubbed a "surveillance society" since the middle of the twentieth century (Brucato 2015). In principle, transparency in a liberal society should be bidirectional: the state surveils citizens to maintain security, while the state's legitimacy is maintained by its accountability and openness to public scrutiny. In practice, however, given the huge asymmetry in resources and expertise at their disposal, there exists an enormous imbalance between how states observe citizens and how citizens observe states.

Ubiquitous connected devices, the avalanche of personal data they collect, and intelligent algorithms to process it have complicated this power struggle. In one sense, the power of states to surveil citizens has never been more evident. Disclosures by Edward Snowden in 2013 depicted a sophisticated surveillance system led by the U.S. National Security Agency. Yet the ubiquity of connected devices in the hands of individuals has also enabled what has been called "sousveillance", or the process of "watching from below" (Mann 2004). This involves observations of the powerful *by* the powerless (or, at least, the less powerful). A vivid example of this phenomenon is the trend towards the video recording of police brutality in the United States, particularly as targeted against African Americans. Starting with the surreptitious recording of Los Angeles police officers beating and kicking the defenceless Rodney King in 1991, "cop watching" has provided a new, technology-enabled way for citizens to observe and scrutinise the actions of law enforcement officials. However, this new ability to witness does not translate automatically into an increase in accountability (Brucato 2015). Police violence in America shows no sign of abating, and a string of acquittals of officers involved in the deaths of suspects—despite

---

[2] It could be argued that the secondary use or reuse of data constitutes an additional stage of the data lifecycle. I offer no contest to this claim, but rather have omitted it here for reasons of space and simplicity.

video evidence appearing to depict their direct involvement—has sparked protest on several occasions (Laughland et al. 2014).

For all these advances in "sousveillance", the infrastructure of data collection and retention remains highly asymmetric. States and large technology corporations enjoy huge apparatus for surveilling individuals, introducing hitherto unseen impositions on the privacy of ordinary people (Schneier 2015; O'Neill 2016).

9.3.2   Distortion at the Point of Data Analysis

As the risk of asymmetry in data infrastructure demonstrates, the increased collection of personal data can have significant consequences. A video clip showing police brutality can provoke a public outcry on its own terms, with only minimal processing. Yet the broader impact of recent technological innovations on transparency and privacy has had more to do with the increasing ability to analyse, process, and make decisions from masses of data than with simply collecting and disseminating it. As the techniques used to analyse this data have become ever more sophisticated—especially with the incorporation of machine learning tools—it has become harder for human operators, let alone subjects, to understand their workings. The lack of transparency around how algorithmic systems work—particularly given the possibility that they will become more effective over time—reduces the ability for individuals subject to their whims to understand or protest their decisions (Wachter et al. 2017a).

Yet the issue is not simply one of a lack of understanding. As well as being coded, algorithms are *en*coded—with the values, beliefs, preferences, and perspectives of their creators. An algorithm inevitably yields biased decisions, reflecting the values of its designer (Mittelstadt et al. 2016). Problems arising in this context range from facial recognition algorithms, which perform far better at recognising Caucasians than African Americans possibly due to the data used to train them (Garvie & Frankle 2016), to Microsoft's artificially intelligent Twitter chat bot, "Tay", which was "trained" to be racist and misogynistic by malevolent (human) Twitter users (Hunt 2016). The Tay affair was a particularly emphatic example of the susceptibility of algorithms to biased training data (see also Diakopoulos 2015). The distortion produced by algorithms, as made possible by bias in the data they are fed, is a systematic reflection of pre-existing values in society more broadly.

Distortion is not always accidental, however, and its effects are not always negative. First, it is important to distinguish between "random" distortion designed to be purely neutral and the more problematic, socially significant distortion that algorithms often display. Randomly-generated differential outcomes are acceptable in various cases. For example, if I buy a lottery ticket that does not win me the jackpot, I have few grounds for legitimate complaint, as long as I am confident the draw was random. Yet a lottery found to favour systematically, say, players with certain demographic characteristics would likely face swift reprimand. Second, distortive techniques can be deliberately utilised to secure a positive outcome, such as improving privacy protection. For example, distortion is the basis of differential privacy, which involves random noise being inserted into samples of data to protect the anonymity of individual subjects.

These counter-examples notwithstanding, data analysis nonetheless offers a potential risk to privacy overall, insofar as the use of analysis may generate new assemblages of individuals with particular characteristics, exposing them to new harms (Taylor et al. 2016).

9.3.3 Discrimination at the Point of Deployment

The third risk involves the differential deployment of insights yielded by data analysis. Whereas the risk of distortion relates to the bias inherent in algorithms and/or the datasets used to train them, the risk of discrimination pertains to the context in which automated decision-making is deployed in the outside world.

This distinction is nuanced but important. Consider examples where an inherently *unbiased* automated decision-making system (to the extent that this is possible to construct in practice) is *deployed* in such a way that its use could generate discrimination. A repressive government, for example, might develop sophisticated facial recognition software, and deploy this on high-resolution photographs of public protests (Singh et al. 2017). Over time, it would be possible to identify protesters who appeared at several rallies, enabling the government to round up the "ring-leaders" of an opposition movement. Note that there is no obvious bias inherent in the algorithm used here—it simply, and accurately, recognises everyone's face—but its repeated deployment in this politically sensitive context will result in significant discrimination, and impositions on the privacy of the keenest participants most of all (Masse 2017).

An analogue equivalent of this case is the policy of stop-and-frisk, wherein police are deployed to specific hotspots of crime, and routinely frisk passers-by for contraband. Even if this process is implemented uniformly on the ground (i.e., police stop and frisk, say, one in every ten passers-by at random), the fact that crime hotspots geographically overlap with concentrations of particular communities— possibly stratified along demographic lines—means that the net effect is an increase in stop-and-frisk for specific groups. In this example, a seemingly rational use of available data (deploying police at crime hotspots) combined with the application of a reasonable and simple "algorithm" (stop and search one in every ten people, no matter their appearance) nonetheless results in an inequitable and unjust outcome. In *Floyd v. City of New York* (2013) a District Court Judge found that in the application of the city's "stop-and-frisk" policy, "those who are routinely subjected to stops [we]re overwhelmingly people of color" (p. 6), in violation of the Equal Protection Clause of the U.S. Constitution.[3] The main difference between the digital

---

[3] Note that in this case, the judge found that the racial composition of a given area predicted the rate at which residents were stopped "above and beyond" the underlying crime rate, making its use disproportionate to an unconstitutional degree. But we argue here that even had there been a perfect

and the analogue examples here, then, is chiefly one of scale and power. An algorithm that could crunch data about the faces of ten thousand protesters in seconds and highlight repeat offenders would likely be exponentially more effective and more immediately actionable than deploying a policy with human "coders" (in the case of stop-and-frisk, police officers).

Deploying automated decision-making might unduly expose given segments of a population, excluding others—even if the algorithm itself can be said to be unbiased. Whether the outcome is positive (e.g., extra benefits) or negative (e.g., extra hassle) for the sub-population targeted, the impact is likely to be significant. This raises the very real prospect that particular segments of a population will find their privacy invaded with increasing regularity as compared with other (more law-abiding? less trouble-making?) citizens.

Framing the discussion in terms of these three risks—asymmetry at the point of collection, distortion at the point of analysis, and discrimination at the point of deployment—enables us to see more clearly the diverse threats that the combination of ubiquitous devices, abundant data and intelligent algorithms pose to privacy. Understanding and overcoming the three risks I have introduced is a necessary first step preventing privacy harms in the contemporary data landscape. In the following section, I present four types of responses which may serve to mitigate these risks.

9.4   A Framework of Responses to Privacy Risks

The diversity of potential privacy risks arising from each stage of the data lifecycle presented in the previous section suggests that an array of responses at multiple levels may be required. In this section I offer a framework designed to be both broad enough to encompass a range of possible responses, and specific enough to offer a concrete path forward. The framework I present is made up of two dimensions.

The first dimension relates to the social level at which the given set of responses applies: the micro level of individuals, or the macro level of society at large. As Wachter et al. (2017a, p. 3) show, algorithmic explanations can be provided at either the level of "system functionality" or "specific decisions". In the broader area of privacy harms that I cover here, a similar distinction between individual- and society-level responses is intuitive. This is because a modern individual is both a data subject themselves, and a member of the wider political community to which they belong. Their interests might reasonably be both in protecting their own privacy rights *viz a viz* the data produced about or by them specifically, and in protecting the privacy of the "average" citizen. (This intuition owes much to Rawlsian social contract theory, in particular Rawls's proposed "veil of ignorance" thought experiment.) I use the category labels "micro" and "macro" as shorthand for this individual-society dichotomy.

---

correlation between stop rate and crime rate (i.e. a perfectly "fair" application of the policy) the burden on the (mostly innocent) people stopped could be considered too severe.

The second dimension of the framework facilitates a distinction between what I dub "instrumental" and "epistemological" responses. There are doubtless circumstances in which individuals may seek to exercise actual control over the flow of information in society, at both the level of their own personal data and at the broader level of society-wide decision-making. Yet other responses to privacy risks take an epistemological form: it is merely enough to understand how, say, a given algorithmic decision was rendered. This notion is central to the "right to explanation" alleged to be contained in the forthcoming General Data Protection Regulation (Goodman and Flaxman 2016; c.f. Wachter et al. 2017a). As will be seen, this epistemological category also explains how extremely sophisticated algorithmic systems could be legitimately used in socially sensitive contexts even when they are beyond the level of public understanding.

These two dimensions together constitute the framework of responses presented below. In the remainder of this section I offer a brief discussion of the significance of each response.

	Instrumental	Epistemological
Micro-level	Control (including consent and contestation)	Comprehension
Macro-level	Consultation	Confidence

9.4.1   Micro-level Instrumental Response: Control

The exercise of individual *control* is perhaps the most intuitive—even reflexive— response to the risks to privacy posed by the flow of personal data. In particular, the notion that people should *consent* to the use of their personal data underlies much of the legal and policy response in this area, particularly as the granularity of personal data, and the profiling this enables, has drastically increased (Hildebrandt 2006). At the other end of the data lifecycle is the idea that individuals should be able to *contest* the outcomes of data analysis when deployed, typically on the grounds that the analysis is in some way flawed (Wachter et al. 2017a). Yet consent and contestation are included here merely as examples of the sorts of approaches that the broader category of control offers.

The idea of controlling or even owning one's own data clearly has appeal, even if these notions, especially data ownership, are uncommon in reality. A recent survey found that the average British consumer values their own private data as being worth £3241, though this far exceeds the market value of private data on the "dark web" (Curtis 2015). Words like "control" and "manage" are found on the privacy settings pages of major platforms like Facebook and Google, fuelling the perception—reality notwithstanding—that users have control over how their valuable data is used. Newer social platforms like Snapchat and WhatsApp have interfaces and protocols that privilege privacy through control (albeit with various caveats). However, the ability for such platforms to keep their services free is chiefly the result of their ability to monetise the very data that users surrender, and this core tension is unlikely to be resolved any time soon. An innovative approach would involve giving users more control at the point of surrendering data, rather than only in retrospect. The Hub Of All Things project, for example, encourages users to think of themselves as controllers of their personal data, and see data as something to be invested or withdrawn from different brokers much like financial capital. Whether in policy or merely in perception, individual control over personal data is evidently a compelling response to the privacy risks introduced by data.

9.4.2   Micro-level Epistemological Response: Comprehension

The ability to confront the specific outcomes of an algorithmic system would be challenging without an understanding of the nature of the problematic decision (Wachter et al. 2017b). This underlies the principle of comprehension, the idea that individuals should be empowered to understand the workings of algorithmic systems, even without necessarily having the capacity to consent to or contest it. This principle clearly has particular significance at the deployment phase of the data lifecycle, where the actual effects of automated decision-making are felt, whether in the context of applying for a job, a loan, or parole (Angwin et al. 2016).

Yet comprehension is not an unalloyed good. Take the "social credit" system in China, whose effectiveness relies in part on citizens having full comprehension of its mechanics. Reports suggest that merely associating oneself socially with a low-ranked fellow citizen can affect one's own "score", in turn affecting the sorts of benefits to which one is entitled (Botsman 2017). Although currently at only a pilot stage, it seems that the program is designed to encourage users to alienate and isolate low-ranked, "undesirable" users, but this mechanism relies on the assumption that citizens know each other's scores, in order to avoid such "undesirables". In this example at least, it is clear that comprehension on the individual level can fuel negative outcomes on a broader scale.

9.4.3   Macro-level Instrumental Response: Consultation

While at the micro-level, individuals might be encouraged or empowered to control, consent to, or contest the misuse of their data, citizens might also play an instrumental role in broader *consultative* processes about how data should be collected, analysed and operationalised. This consultation might take the form of public engagement around specific laws, policies or practices, or about the design of specific technical systems. An example of the latter case is the Moral Machine project based at MIT's Media Lab, which presents participants with a series of iterations of the "trolley problem" reimagined for self-driving cars, forcing unpleasant choices as to which way a car should swerve (MIT 2017). A broader consultative effort is currently being undertaken by Royal Society of the Arts in association with DeepMind, who have proposed a "citizen's jury" on the use of AI in criminal justice (Balaram 2017).

Consultative initiatives like the Moral Machine project and the citizen's jury are an essential part of the broader set of responses to privacy risks in the digital age, and apply at every stage of the data lifecycle. The emergence of AI systems in the public sphere means that not every individual potentially affected by automated decision making will be able to consent to it on an individual level; consider, for example, the cyclist in the cross-hairs of a self-driving car. In these cases, everyone should in principle be consulted (or at least "consultable") about how automated decision-making systems should be designed and their outcomes operationalised.

9.4.4 Macro-level Epistemological Response: Confidence

Taken together, the three categories of responses just presented capture almost all possible modes of privacy protection. Yet as we have seen, each also has flaws. The idea of individuals controlling or even owning their data conflicts with the basic model of data-driven advertising that undergirds the preeminent business model of the modern technology sector. People understanding how their data is being used seems at first glance eminently sensible, but as the Chinese social credit example demonstrates, too much knowledge can have a powerful effect, and perhaps even be deleterious to the fabric of society. Consultative efforts seem useful, but it remains to be shown that these can work on a societal scale.

Even putting these limitations to one side, it is arguable that not every AI system could or should be subject to individual control, comprehension or public consultation. It is important to acknowledge situations where, whether for reasons of public safety or economic security, individuals should not be in a position to control, comprehend or even be consulted on data collection, analysis, or use. Consider, for example, the use of algorithmic techniques to detect large-scale fraud. Financial institutions have developed sophisticated algorithms, based on highly sensitive data, to estimate the likelihood that a given transaction was fraudulent. Or consider the sort of complex systems that we might assume exist to predict terrorist activity. In these and other situations, it is neither plausible nor desirable for the individuals under investigation to have control over how data about them is being used, to understand how it is being used, or even—at least in the most sensitive of cases—for society at large to be consulted as part of the designs of such systems.

For this very small set of examples, the only type of response may be mere "confidence" in the fact that, where such privacy-invading activities are taking place, they are being done ethically. This is sub-optimal by the standards of almost any conception of liberal democracy, and is perhaps closest in spirit to Thomas Hobbes's justification of an all-powerful "Leviathan" to protect the public from external threats (2006[1651]). Yet however reductive on contemporary ethical grounds, it bears acknowledging the circumstances in which mere confidence in a system is the only recourse, such as it is, to the risks to personal privacy of data collection, analysis and operationalisation. Public confidence is something that is won only over generations and can be lost overnight. But while it holds, it can serve to provide a very thin form of legitimacy for the capture, use, and analysis of personal data in the absence of more robust protective measures.

The framework introduced in this section is merely a first systematic step towards what I hope will be a concerted effort to understand the variety of possible responses to the unprecedented risks to privacy emerging in the age of ubiquitous devices, abundant data, and sophisticated, automated analysis. By categorising and organising these responses in the way presented, it is possible to develop a fuller sense of the level at which recourse to privacy threats may be applied, and the respective value of distinct instrumental and epistemological approaches.

9.5   Conclusion

In this chapter I have argued that privacy as a right has long been tightly interwoven with the technology of the time. The ability to capture information of value—with the threat to privacy that this involves—relies on technology of some kind to do so. This is as true for the most basic types of technology, such as ink and paper, as it is for far more modern forms of data collection, such as CCTV cameras and smartphones. In recent decades, however, radically more sophisticated ways of analysing this data, and operationalising the insights that this analysis offers, have led to proportionally increased threats to personal privacy, at each stage of the lifecycle of data. This includes the risk factors of asymmetric infrastructure for the collection of data, systematic distortion in the analysis of data, and the discriminatory ways that the insights of algorithmic systems are deployed. The scale and scope of these risks to privacy demand a multimodal set of responses, and the framework I developed in this chapter represents an attempt to collect and categorise these responses in relation to the level at which they apply and the mode in which they operate. This framework may therefore serve as a means of organising and coordinating the raft of responses that are required to ensure that privacy remains a respected right, in principle and in practice.

References

Angwin, J., J. Larson, S. Mattu, and L. Kirchner. 2016, May 23. *Machine bias.* ProPublica. Retrieved January 13, 2018, from https://www.propublica.org/article/machine-bias-risk-assessments-in-criminal-sentencing

Balaram, B. 2017, October 4. *The role of citizens in developing ethical AI.* The Royal Society for the Arts. Accessible at: https://www.thersa.org/discover/publications-and-articles/rsa-blogs/2017/10/the-role-of-citizens-in-developing-ethical-ai

Botsman, Rachel. 2017, October, 21. *Big data meets big brother as China moves to rate its citizens. Wired.* Accessible at: https://www.wired.co.uk/article/chinese-government-social-credit-score-privacy-invasion

Brandeis, L.D. 2009. *Other people's money and how the bankers use it.* Cosimo.

Brucato, B. 2015. The new transparency: Police violence in the context of ubiquitous surveillance. *Media and Communication* 3 (3).

Curtis, Sophie. 2015, November, 23. *How much is your personal data worth? The telegraph.* Accessible at: https://www.telegraph.co.uk/technology/news/12012191/How-much-is-your-personal-data-worth.html

Diakopoulos, N. 2015. Algorithmic accountability: Journalistic investigation of computational power structures. *Digital Journalism* 3 (3): 398–415.

Floridi, L. 2005. The ontological interpretation of informational privacy. *Ethics and Information Technology* 7 (4): 185–200.

———. 2016. On human dignity as a foundation for the right to privacy. *Philosophy & Technology* 29 (4): 307–312.

———. 2017. Digital's cleaving power and its consequences. *Philosophy & Technology* 30 (2): 123–129.

Floyd v. City of New York, 959 F. Supp. 2d 540 (S.D.N.Y. 2013).

Garvie, C., and J. Frankle. 2016. Facial-recognition software might have a racial bias problem. *The Atlantic.* Accessible at https://www.theatlantic.com/technology/archive/2016/04/the-underlying-bias-of-facial-recognition-systems/476991/

Glancy, D.J. 1979. Invention of the right to privacy. *The Arizona Law Review* 21: 1.

Gibbs, Samuel. 2018, March, 7. *Google's AI is being used by US military drone programme. The Guardian.* Accessible at: https://www.theguardian.com/technology/2018/mar/07/google-ai-us-department-of-defensemilitary-drone-project-maven-tensorflow

Goodman, B., and S. Flaxman. 2016, June. EU regulations on algorithmic decision-making and a "right to explanation". In *ICML Workshop on Human Interpretability in machine learning (WHI 2016),* New York. Accessible at http://arxiv.org/abs/1606.08813.

Hildebrandt, Mireille. 2006. Privacy and identity. In *Privacy and the criminal law,* ed. Erik Claes, Antony Duff, and Serge Gutwirth. Oxford: Intersentia.

Hobbes, T. 2006 [1651]. *Leviathan.* A&C Black.

Hunt, E. 2016. Tay, Microsoft's AI chatbot, gets a crash course in racism from Twitter. *The Guardian.* Accessible at https://www.theguardian.com/technology/2016/mar/24/tay-microsofts-ai-chatbot-gets-a-crash-course-in-racism-from-twitter

Katz v. United States, 389 U.S. 347 (1967).

Laughland, O., K. Epstein, and J. Glenza. 2014, December 5. *The Guardian.* Accessible at https://www.theguardian.com/us-news/2014/dec/05/eric-garner-case-new-york-protests-continue-through-second-night

Lomas, N. 2017. Google's right to be forgotten appeal heading to Europe's top court. *TechCrunch.* Accessible at https://techcrunch.com/2017/07/19/googles-right-to-be-forgotten-appeal-heading-to-europes-top-court/

Mann, S. 2004, October. Sousveillance: Inverse surveillance in multimedia imaging. In *Proceedings of the 12th annual ACM International conference on multimedia,* 620–627. ACM.

Masse, B. 2017, Sepetember 12. What's the worst that could happen with huge databases of facial biometric data? *Gizmodo*. Accessible at https://www.gizmodo.com.au/2017/09/whats-the-worst-that-could-happen-with-huge-databases-of-facial-biometric-data/

MIT. 2017. *The moral machine*. Accessible at: http://moralmachine.mit.edu

Mittelstadt, B.D., P. Allo, M. Taddeo, S. Wachter, and L. Floridi. 2016. The ethics of algorithms: Mapping the debate. *Big Data & Society* 3 (2): 2053951716679679.

Nakashima, E. 2016. Apple vows to resist FBI demand to crack iPhone linked to San Bernardino attacks. *Washington Post*. Accessible at https://www.washingtonpost.com/world/national-security/us-wants-apple-to-help-unlock-iphone-used-by-san-bernardino-shooter/2016/02/16/69b903ee-d4d9-11e5-9823-02b905009f99_story.html

New York Times Co. v. United States, 403 U.S. 713 (1971).

Nissenbaum, H. 2009. *Privacy in context: Technology, policy, and the integrity of social life*. Stanford: Stanford University Press.

O'Neill, C. 2016. *Weapons of math destruction. How big data increases inequality and threatens democracy*. New York: Random House Audio.

Olmstead v. United States, 277 U.S. 438 (1928).

Panday. 2017, August 28. India's Supreme Court upholds right to privacy as a fundamental right – and it's about time. *Electronic Frontier Foundation*. Accessible at https://www.eff.org/deeplinks/2017/08/indias-supreme-court-upholds-right-privacy-fundamental-right-and-its-about-time

Posner, R.A. 1978. Economic theory of privacy. *Regulation* 2: 19.

Puttaswamy v. Union of India. 2017. Writ Petition (Civil) No. 494 of 2012.

Schneier, B. 2015. *Data and goliath: The hidden battles to collect your data and control your world*. New York: WW Norton.

Singh, A., D. Patil, G.M. Reddy, and S.N. Omkar. 2017. *Disguised Face Identification (DFI) with facial keypoints using spatial fusion convolutional network*. arXiv:1708.09317v1 [cs.CV].

Solove, D.J. 2005. A taxonomy of privacy. *University of Pennsylvania Law Review* 154: 477.

Taylor, L., L. Floridi, and B. van der Sloot, eds. 2016. *Group privacy: New challenges of data technologies*. Cham: Springer.

Thomson, J.J. 1975. The right to privacy. *Philosophy and Public Affairs* 4 (4): 295–314.

Wachter, S., B. Mittelstadt, and L. Floridi. 2017a. Transparent, explainable, and accountable AI for robotics. *Science robotics* 2 (6): eaan6080.

Wachter, S., B. Mittelstadt, and C. Russell. 2017b. *Counterfactual explanations without opening the black box: Automated decisions and the GDPR*. arXiv:1711.00399.

Warren, S.D., and L.D. Brandeis. 1890. The right to privacy. *Harvard Law Review* 4: 193–220.

Chapter 10
Digitalised Legal Information: Towards a New Publication Model

Václav Janeček

Abstract This chapter outlines key developments regarding publication and communication of legal rules and standards (i.e. legal information) to show that dissemination of legal information is reliant on how we design the entire model of its publication. In doing so, it analyses paradigmatic models of publication as they appeared in the prehistorical, historical, and hyperhistorical stages of human evolution. These models demonstrate how legal information was delivered to its intended addressees, i.e. to those who were expected to obey the published laws. It also demonstrates that the progress regarding these publication models was driven by efficiency and sustainability considerations. The currently prevailing model of publication is, however, inefficient and unsustainable due to an unnecessary multiplication of intermediaries facilitating communication of legal information. This problem is even more apparent in the context of increasing digitalisation of legal information and emerging information and communication technologies (ICTs). The chapter argues that, in this light, it is appropriate to consider revising the entire publication model and not only some aspects of it. An addressee-centric publication model is outlined as a potential solution to the problem. The proposed model requires active delivery of a relevant subset of digitalised legal information to its intended addressee in a similar way as targeted online advertising. Unlike the existing research that promotes personalisation of law (personalised legal information), this chapter advocates personalisation of the publication model.

10.1 Introduction

Do you know how many new legal acts or regulations begin to apply to you each year? If you lived in the European Union (EU) in 2017, for example, it would be up to 131 entirely new directly applicable regulations and up to another 230 regulations

V. Janeček (✉)
Faculty of Law and St Edmund Hall, University of Oxford, Oxford, UK
e-mail: vaclav.janecek@law.ox.ac.uk

© Springer Nature Switzerland AG 2019
C. Öhman, D. Watson (eds.), *The 2018 Yearbook of the Digital Ethics Lab*,
Digital Ethics Lab Yearbook, https://doi.org/10.1007/978-3-030-17152-0_10

amending the already existing EU law in force (EUR-Lex 2018). This means that, on average, approximately one new regulation was enacted each day last year. Suffice to add that apart from regulations there are many other types of legal sources that contain enforceable laws. Now, although these EU laws are freely available via the EUR-Lex online search engine, one has significant doubts about the size of the group out of the EU28's 511.8 million total population (Eurostat 2018) that has even the slightest idea about what these regulations regulate. As a lawyer myself, for instance, I know only a fraction of them.

A simple question arises then as to how addressees of the EU regulations can comply with such laws if they do not know their content or are not even aware of their existence in the first place. One option for them is to pay massive amounts to legal specialists and thereby gain sufficient legal knowledge to ensure necessary compliance with the laws. This setting, however, is inefficient and unsustainable because members of society are pushed towards desirable behaviour by lawyers holding an imaginary carrot and stick, rather than being sustainably educated and informed about relevant legal rules and standards (i.e. legal information). EU regulations are only rarely discussed publicly—such as in the recent case of the General Data Protection Regulation (GDPR)[1]—and our society thus often remains legally incompetent, uninformed or even misinformed.

Had the lawyers and other legal specialists taken away the imaginary carrot and stick, most of us would soon be oblivious to the officially published legal information. This, in my view, is a problem because our society usually does not internalise newly published legal rules and standards and, therefore, does not mature legally or matures only very slowly. In this chapter, I argue that this does not need to be the case if we change the model of publication of digitalised legal information.

Imagine living in year 2096. Oxford University celebrates its 1000th anniversary, the Digital Ethics Lab publishes the 79th volume of its yearbook, and you—being part of a legally mature information society where peoples' expectations regarding legal rules almost perfectly match the existing regulations—enjoy living a decent life. In this not-too-distant future, a significant amount of digitalised legal information is directly implemented into software code and executed without human intervention (see e.g. a list of start-ups at Legal Geek 2018). The artificial intelligence (AI) based technological milieu largely eliminates human intermediaries to avoid inefficiencies, thus leaving the environment free of informational ballast that no human individual could comprehend or process. The remainder of legal information, i.e. non-self-executing information, is published under such a model that allows every relevant piece of legal information to be delivered to its addressee and to publicise and communicate this information repeatedly in appropriate contexts, thus efficiently and sustainably educating this addressee and, in turn, helping our society to mature legally.

---

[1] Regulation (EU) 2016/679 of the European Parliament and of the Council of 27 April 2016 on the protection of natural persons with regard to the processing of personal data and on the free movement of such data, and repealing Directive 95/46/EC (General Data Protection Regulation) (OJ L 119/2016).

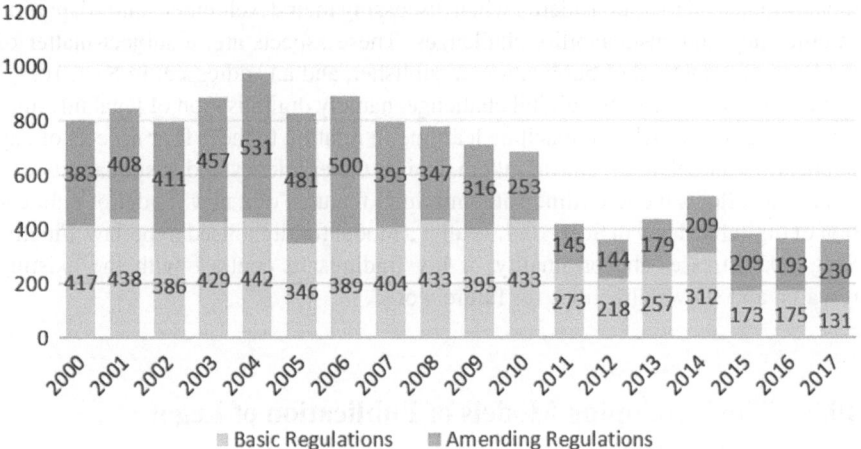

Fig. 10.1 Number of EU law Regulations enacted by year and type. (Data source: EUR-Lex 2018)

Existing models of publication of legal information (at least in Europe) are yet far from the futuristic scenario as set out above, which gives us room for improvement. The current model of publication of EU law, for example, is a static one: the EU legal rules are produced in the EU legislative bodies and published in a static journal that is available online via the EUR-Lex search engine: the e-OJ (electronic Official Journal of the European Union). EU citizens who want to find legal information freely must access actively this static journal themselves and perform a qualified search task. Alternatively, they can set up an alert system that will feed them with information about myriads of new EU legal documents. Therefore, although the EU regulations statically wait to be discovered and read by the law-abiding citizens, their quantity makes it unrealistic for the readers to actually get to know them. If the readers were to learn all the legal sources via this Web 1.0 interface, they would have no time left for doing anything else with their lives than reading said regulations (see Fig. 10.1). For non-specialists, searching for legal information thus inevitably becomes merely a heuristic task aimed at solving an individual problem. They are only interested in finding the relevant EU law when they deal with their own issues that they perceive as a legal problem. Accordingly, it is fair to assume that the EU citizens care nothing for the EU regulations at large, abstract knowledge of these regulations, and legal maturity of the society. The point is that, in a long run, such model is inefficient and unsustainable.

In this chapter, I address the problem of publication models by exploring how these models evolved. In Sect. 10.2, I thus look at various stages of their past developments and use some iconic examples to demonstrate how models of publication of legal information are intertwined with the developments of information and communication technologies (ICTs) and how these models addressed the problems of inefficiency and unsustainability. Then I analyse common aspects of all publication models that need to be considered when discussing their development and adaptation to efficiency and sustainability challenges. These aspects are: a subject-matter of publication, a method of publication, a publisher, and an addressee. In Sect. 10.3, I explore the specifics of the digital challenge, namely digitalisation of legal information, emergence of AI, and machine learning in relation to these four aspects of any publication model, and I canvass them against the efficiency and sustainability criteria. This allows me to outline some minimal features of a new model of publication of digitalised law in Sect. 10.4. I call it a model for digitalised hyperlaw. Finally, Sect. 10.5 stresses the originality of my findings in contrast with the existing research and shows directions for future work.

10.2   The Developing Models of Publication of Legal Information

Looking back through the evolution of human societies, we can identify three distinct stages: prehistory, history, and hyperhistory (Floridi 2012). On this account, prehistory can be regarded as an era in which information is not recorded due to lack of recording methods. With Floridi we can thus describe prehistory simply as a stage in societal evolution before "the invention and development of information and communication technologies (ICTs)" (Floridi 2012: 129). The historical stage, then, "began with the invention of writing in the fourth millennium BC" (ibid: 129) which made it possible to record and transmit data over time and thereby create historical information. Finally, the hyperhistorical era marks a significant change in which societies use ICTs not only to record and transmit data but also to process them, which makes such hyperhistorical societies "vitally dependent on [… ICTs] and on information as a fundamental resource" (ibid: 130). It is important to mention that the developments in ICTs are not global. On one hand, some present societies live prehistorically, i.e. without ICTs, on the other hand the EU finds itself only at the onset of hyperhistory, because large parts of EU are still not reliant on ICTs processing of data and legal information.

In this context, recording, transmitting, and processing of data and legal information play a distinct role in each of these three societal stages and, as I will outline in this section, the emergence and developments of ICTs shaped significantly the models of publication of legal information. Accordingly, I will show that the developments of ICTs had a major impact on how societies legally organised themselves. In doing so, I will pick some paradigmatic models of publication from each of the three stages of societal evolution to demonstrate that the progress regarding publication models was driven by efficiency and sustainability considerations. We will see that each new model of publication and communication of legal information was expected to be more efficient and sustainable that the previous one.

10.2.1   Prehistorical Models

Since long before people started recording legal information in writing, they used symbolic gestures and pronounced solemn words upon legally important transactions such as when concluding a contract or entering a marriage. They did it to mark the importance of the moment and to create at least an impression of the existing legal bond (Charpin 2010: ch 3). In such context, it was the witnesses in front of whom these gestures were performed and who thus testified the existence of such legal bond. In this sense, we can say that the witnesses acted as human recorders of relevant legal information, which made it slightly problematic in an environment where people could die easily and where life expectancy was very low. Legal information was thus only a short-term type of information and, on the top of it, it was not very transparent. Similarly, any superior societal authority, i.e. a powerful leader of a social group, who wanted to broadcast her or his own legal information could only rely on this inefficient model. So, for example, when someone got authoritatively punished for trespassing other's assets (e.g. for killing or stealing his life stock), the information about impermissibility of such trespass was only disseminated via stories and via vivid experiences of those involved in the execution of the (usually very harsh and painful) physical punishment.

In the context of such unwritten, short-term, and highly volatile system of recording, transmitting and processing of legal information, the rise and implementation of writing as a new technique presented an immense progress. Hence, written records of legal rules and legal information were soon demanded as the new standard. In fact, "a few jurists have believed that paragraphs from the Code of Hammurabi express the obligation imposed by the king to fix in written form contracts (relating to marriage, herding, or tenant farming), or risk having them invalidated" (ibid: 48). Not only did the invention of writing enable long-lasting access to legal information—e.g. by using clay tablets to record debts, transfers, and property claims in Old Babylonian kingdom (ibid: 49)—but it also facilitated a more structured and centralised model of publication of legal information. The sovereign could publish all the relevant legal information by a uniform method that was in principle independent of the mortal human "recorders" and "interpreters". One such method (to which I will turn in the next section) was to set legal rules in stone pillars that were erected in public places.

The dividing line between prehistory and history was never crystal clear because prehistory and history, "prelaw and law, or world of rite and oath on one hand, world of writing on the other" (ibid: 51) existed in parallel. These two eras were thus inextricably mixed, as for example Babylonian laws from the twentieth to seventeenth century BC demonstrate (ibid: 51). Accordingly, to speak of a prehistorical model of publication of legal information, we need to focus on its ahistorical features. For the purposes of this chapter, therefore, the prehistorical model could be best described as a model of shared life experience, a model that could only accommodate and disseminate small amounts of legal information to a small number of people in a limited geographical area and for a short period of time. As such, the prehistorical model was generally inefficient and unsustainable, especially if we take a maturing society ruled by law as our benchmark.

10.2.2   Historical Models

Some of the oldest written statues, and therefore evidence of some of the oldest historical models of publication of legal information are the famous Babylonian Code of Hammurabi (1754 BC) and Sumerian Code of Ur-Nammu (c. 2100–2050 BC). The Sumerian legal text is still considered the oldest surviving code on the planet (Kramer 1958: ch 7). What is typical of these codes is that they were carved into persistent materials such as stone and were stored and displayed at prominent places in the kingdom so that every subject to the king could come and see the binding law (i.e. key legal information) himself or herself. Practically the same model using physically immovable or only hardly moveable materials as a recorder of legal information was used in the Law of Twelve Desks (*Leges Duodecim Tabularum*) in the early Roman Republic (449 BC). Efficiency of this model was based on the fact that people went to public places where these laws were permanently displayed and where everyone could thus learn them. In comparison with the prelaw period, historical legal information could have been more easily preserved over generations and in principle, everyone could have gotten familiar with them. This did not apply only to authoritative rules but also to rules laid down in written contracts.

Still in the historical era, the advancements of ICTs—for instance the invention and implementation of transportable information carriers such as papyrus, parchment, or paper—made it easier to transform the static model of displayed information into a model where the promulgated laws could have been copied (rewritten) and disseminated across the territory in which they were supposed to be binding. Gradually, handwriting on lighter carriers became more widespread and made it possible to disseminate legal information widely. This could be seen as an advantage but also as a problem because legal information was communicated and disseminated rather chaotically, without a clear publication blueprint. The model was becoming less centralised and although common laws could have been transferred from one place to another simply by transporting the relevant legal document, the local laws could have evolved, and often also did evolve, in substantially different ways.

One way of dealing with this piecemeal publication model was to collect existing historical laws as they developed locally and provide a comprehensive overview of them. This was, for example, the case of *Corpus Iuris Civlis* which was compiled in the sixth century AD subject to order by Eastern Roman Emperor Justinian I, and was then distributed across the Empire as authoritative evidence of existing laws. We can interpret this initiative as an attempt to re-gain control over legal information in the Emperor's territory, and also to monopolise the publication of such information by identifying an authoritative source of that information.

Another interesting model appeared in medieval England, shortly before one such formal source of legal information called *Magna Carta Libertatum* (1215 AD) was agreed to by King John of England. The English model addressed the problem of decentralisation not by collecting and redistributing common (yet locally distinct) laws, but by implementing a specific system of justice. The efficient dissemination of legal information was facilitated by travelling "itinerant" judges who dispersed justice according to common laws by going from town to town and hearing disputes of individuals. Paradigmatically, the model thus no longer required the royal subjects to travel to London to find out what the law was. The laws (figuratively speaking) travelled with the judges to the king's subordinates. In a sense, this model was very similar to the one of Eastern Roman Empire, except that it was not an *a priori* written collection of rules, but an *a posteriori* deciding judge who travelled around the country.

Even though it was the official authorities who were responsible for the "travels" of legal information via the court system and therefore could have maintained control over information dissemination, this model was bound to fail at keeping all published common laws comprehensible. This failure was due to both the increasing amount of judgments, as well as the practical *a posteriori* orientation of the publication model which was not focused on gathering comprehensive information about the common law system, but mainly on applying and disseminating particular legal information as widely as possible. No one could have, at the time, followed the work of every single judge across England, and therefore no one could have known how the common law was evolving. In practice, thus, there was a shortage of access to the increasing body of common law. A provisional solution to this problem was then similar to the model of the Eastern Roman Empire: Sir William Blackstone (eighteenth century AD) collected the existing laws and gathered them into systematically structured volumes entitled Commentaries on the Laws of England. These volumes cleared the way for more principled and comprehensive recording and transmitting of laws. The missing *a priori* piece in the existing *a posteriori* English publication model was thus found.

In the days of Blackstone, one very efficient form of ICT was already in place: the printing press. Historical estimates show that the overall

> European book production increased enormously [thanks to advancements in printing technologies] from somewhat more than 12,000 manuscripts per century (or 120 per year) from 500 to 700, to more than one billion books published during the eighteenth century (the peak year in the period 500–1799 is 1790, when more than 20 million copies were printed). (Buringh and Van Zanden 2009: 417)

This made it possible to think anew about how to record and publicise legal information, even though books and printed manuscripts were still regarded as luxury goods (ibid: 440). The potential for developing a new publication model must have been obvious to anyone in the business of publishing.

The development of legal information and its publication models could have taken multiple routes, but a historical coincidence had it that, at the onset of the nineteenth century, Europe witnessed an important codification movement resulting in the *Code civil des Français* (Code Napoléon), the Austrian *Allgemeines bürgerli-ches Gesetzbuch* (ABGB), etc. The preceding advancements in printing and overall increased literacy in Europe (Eisenstein 2012: ch 4) then supported the idea that laws could be written down in a compact form—a codification—to be distributed and read widely, much like the Bible. And although positive laws, unlike the Bible, proved to be constantly changing, this historical period made a significant step towards a uniform top-down model of printed publication which was very efficient and which still underpins many of the recent models of publication. The newly gained advantage was that if there were any amendments to the original legal text, the legislator could have only issued the amendment and the addressee could then physically attach this amending piece of information to the designated page in his or her own copy of the codification. Such model allowed for an unprecedentedly wide and efficient dissemination of legal information. The model was, on the face of it, also sustainable.

However, at the same time when the historical publication model started adher-ing to the idea of codifications, printing was still a low scale business. For instance, the Czech translation of the ABGB from 1812 was never printed in sufficient num-bers so that Czech citizens could learn the civil code (ABGB) properly. Evidence suggests that printed translations of the ABGB were scarce and their copies quickly sold out (Janeček 2017: 52).

This was a problem only until mid-nineteenth century, though, when a much more productive rotary printing press appeared. The rotary printing press made things easier but, at the same time, it spurred new troubles. On the one hand, the rotary printing technology in combination with the rise of industrial revolution steeply increased the capacity of printing houses in comparison with those using the outdated flat printing technology. As a result, legislators could now deliver printed legal information to virtually everyone in a relatively short time. The scarcity of printed legal information was no longer an issue. On the other hand, the invention of rotary printing press appeared too late to affect the already existing codification-based as well as printing-press-reliant publication model. It only changed one aspect of this model, namely the method or technology of publication.

One small step for publishers, but one giant leap for legislators. That is how we can describe the change that took place in the second half of the nineteenth century. The new rotary printing technology, this relatively small advancement of publish-ers' capacity successfully defeated one important presumption that was present in every legislator's publication model until then. The axiomatic presumption was that laws cannot be changed instantly because communication technologies would not allow for efficient publication of such changes. Towards the end of nineteenth cen-tury, however, this axiom was no longer valid. The development of ICTs (i.e. rotary printing technology) removed factual constraints on how frequently legislative changes and new rules can be made and, as we know from history, our law-making authorities took full advantage of this shift. This was the giant leap for legislators.

After this change, new rules and amendments are produced like never before, the rule-making increases and leads to what we may now call a hyperhistorical model of publication of legal information. In the upcoming models we must rely on ICTs not only to record and transmit legal information, but also process it, for otherwise we would not be able to navigate ourselves through this emerging "hyperlaw" and to manage the enormous amount of legal information.

10.2.3   Hyperhistorical Models

To understand how hyperlaw has emerged, consider this outrageously simplified analogy. Imagine a Christmas tree. That is the initial body of law which was published by the rotary printing press regarding one specific area of law, for example the law of contract. Next, imagine that you start decorating the tree. The decorations represent all the new rules, amendments, and new cases in contract law which change the overall picture of this domain. Now, although everyone can buy such Christmas tree and although everyone can be equipped with the newest edition of the legal decorations, not everyone would know where to place these decorations on the tree. In the first instance, you could solve this problem by paying a specialist who will decorate the tree on your behalf, so that you can still see it as an intelligible source of legal information—as a nice Christmas tree. Had there been only one Christmas tree for every household, this solution could be, at least in theory, sustainable. But apart from the contract law pine tree, you would also need to have a tort law spruce tree, a criminal law fir tree, and so on. It is thus only feasible for a specialist to keep all the trees properly decorated. That is the existing model. It only allows the addressees of legal rules to come and see the tree for some ridiculously high fee that they pay to a specialist.

At this stage, complex legal information—the full array of Christmas trees—is privatised and *de facto* not freely accessible. A further problem is that once you create a market in Christmas decorations, you also incentivise production of new decoration sets, and the trees then start getting bigger and bigger. The growing piles of new decorations will eventually turn the Christmas tree arboretum into a messy jungle with no clear boundaries. At that point, what you get for your viewing ticket is anything but a series of "maybes" and unclear interpretations of the smudged images of the presumed body of law. Obviously, such model is not only inefficient and unsustainable, but also extremely costly. To repair this unappealing picture, lawyers first came up with a more specialist education of the professional decorators to train them to absorb the arboretum growth. Then they came up with creating legends, indexes and further guides to legal information in order to retain the disappearing picture of individual Christmas trees. Finally, we reached a point when we started outsourcing the processing of legal information (decorating of the trees) to new ICTs and their operators. This was the onset of hyperlaw and it started with the first computers, even before the dawn of the Internet (Bench-Capon et al. 2012; Leith 2016). Hyperlaw and hyperhistorical model thus arose in response to the unsustainable growth of legal information. Rather than to look at the jungle with awe, we started looking at it with new ICTs.

Currently, the hyperhistorical model looks roughly as follows. The formal sources of legal information are gathered by intermediaries controlling new ICTs. These ICTs are usually referred to as "legal information systems" (see Biasiotti and Faro 2011).[2] They facilitate easier access to legal information by providing automated search engines and providing computing power that vastly outperforms humans in decorating the imaginary Christmas trees. This model makes it feasible to keep track of published legal information, but it demands everyone to go actively to the ICT-run database, just like everyone had to go to public places in the Sumer empire to see the legal engravings in stone pillars. The efficiency and sustainability of the hyperhistorical model is also undermined by the fact that most of existing legal information systems are not freely accessible.

10.2.4   Four Aspects of Any Publication Model

We have seen that changes to the models of publication of legal information correlate with changes to the ICTs, and that these changes seem to respond to the problems of efficiency and sustainability of the models of publication. Perhaps there is even a causal link between the said issues. Here it suffices to say that such causation would not be unidirectional in a sense that the changing ICTs or the efficiency and sustainability problems would cause the changes to the model of publication. Instead, the causal link goes both ways. The examples show that ICTs can both cause and solve efficiency and sustainability problems, and that a new model can also eventually lead to these problems. To unfold these issues, it is useful to divide conceptually any publication model into minor parts and to show how these parts relate to ICTs and problems of efficiency and sustainability.

Looking at the three types of model (prehistorical, historical, and hyperhistorical), we can say that each publication model must entail four aspects: a subject-matter of publication; a method of publication; a publisher of legal information; and an addressee of published legal information. In this fourfold classification, the invention of the printing press, much like the invention of writing, internet, and other ICTs is best described in terms of a publication method. In themselves, these ICTs tell us little about what, by whom, and to whom legal information is published in a given model. A publication model, in contrast, represents a holistic description of a system in relation to which we can better assess efficiency and sustainability of such model. Take for instance the Sumerian method of erecting massive stone blocks with the code engraved on it. Such method may and may not be considered efficient and sustainable only when considered in relation to remaining aspects of a model.

---

[2] See Free Access to Law Movement. http://www.fatlm.org/ [https://perma.cc/3SMF-KZGH]; World Legal Information Institute. http://www.worldlii.org/ [https://perma.cc/322G-BMSD]; Legal Information Management journal—ISSN: 1472-6696 (Print), 1741-2021 (Online).

10.3   The Digital Challenge, AI, and Machine Learning

Most recent developments in ICTs present "a digital challenge". This includes digitalisation of legal information, the emergence of the so-called intelligent ICTs popularly referred to as AI, and the increasing implementation of machine learning algorithms in processing of (but not only of) legal information. The digital challenge now poses a big obstacle to the existing model of publication. At the same time, it poses an opportunity for a significant improvement. Today's ICTs allow us to do things and process information in a way that was not only impossible but also unthinkable a decade ago. The next big question is thus how the existing publication models could accommodate these new ICTs.

It seems to me that there are two major alternatives how we may address the digital challenge and how we may use these new "smart" ICTs. On the one hand, it could result in implementing better *methods* of publication within the existing "Christmas tree" model as outlined in Sect. 10.2.3. On the other hand, it could be seen as a challenge to the model as such. Let us therefore consider implications of the digital challenge in relation to all four aspects of any publication model (a subject-matter; a method; a publisher; an addressee) to see whether the second variant could be justified in the light of the challenge.

10.3.1   A Subject-Matter

The most apparent implication of the digital challenge for the subject-matter of legal information is twofold. For one, legal information has been digitised and practically all positive laws in Europe now formally (also) exist in some form of a digital sequence. This does not in itself alter the substance of such information, i.e. its meaning. Yet the machine-readable digital form of legal data further magnifies our dependence on ICTs, for without ICTs we would no longer understand the digital data set. And although the subject-matter still coexists in digital (machine-readable) and analogue (human-readable) from, the digital challenge slowly puts more weight on the digital form of legal information. For instance, if a text of a statute is amended by the legislator, then the consolidated *full-text* of the amended statute is typically produced only in the digital form. Derivatively, the ICTs extract information from the digital data representing the amended statute so that we can read it as a full-text. The first implication thus relates to the formal expression of legal information. A danger in this regard is that digital data will overtake the role of a primary source of legal information, rather than secondary formal expression of such information. Such scenario is not unlikely to happen given that expert legal discourse often confuses data with information (Janeček 2018).

The second implication applies to the substance of legal information. The digital challenge affects "what" legal information is recorded, transmitted, and processed. On one hand, there are new ICTs for existing laws, such as the emerging phenomenon of smart contracts. Smart contracts are designed to process contractual obligations without human intervention, i.e. automatically (Cuccuru 2017). This eliminates several steps in the process of application of contractual law, which is cost efficient. Yet it also black-boxes legal information that are processed during those steps. The contractual parties no longer need to understand the rules and legal information that are applied by the smart contract. They merely see the result—usually an information about legal sanction (e.g. contractual penalty or tax deduction) that was automatically applied by the smart contract. As a result, this could undermine the role of primary legal information (rules on how one should act) and stress the role of secondary sanctions, thereby eliminating certain types of legal information from the discourse and from publication. Whether this is a good thing is something that is still to be explored, albeit not in this chapter.

On the other hand, we may also think of the digital challenge as giving rise to substantive laws for the new ICTs. In other words, we can contrast ICTs for law with law for ICTs (cf Ruhl 2018). Debates on laws for new ICTs, AI, and for machine learning algorithms are now thriving in Europe and beyond (e.g. Lepri et al. 2017; Wachter et al. 2017). It thus proves important to consider, inter alia, to whom these regulations should be addressed and whether, or to what extent, such information should be seen as part of some unitary legal information category. I will return to this issue in Sect. 10.3.4, but here it is important to mention that new regulations for new ICTs need not to be instructional in the sense as a rule not to kill people is instructional. Instead, they can take the shape of organisational norms or standards just like the EU regulation of electronic signatures (signatures that fulfil certain requirements will gain the same level of legal protection in all EU Member States).[3] If that is the case, then efficiency of publication of such information does not need to be measured by how many addressees know about such standard, but also by how many new ICTs comply with such standard. In other words, in light of the digital challenge, the entire publication model could be evaluated against the direct implementation of legal information.

10.3.2 A Method

As regards the method of publication of legal information, the emerging technological advancements put forward new opportunities for how to record, transmit, and process legal information, as well as how to publish it (see Bench-Capon et al. 2012; Leith 2016). The vast majority of implementations of new ICTs try to enhance the existing models by employing the computational power of ICTs and therefore processing of more data and legal information. This means that most of new solutions, as far as one can tell by literature review or by looking at start-ups in law &

---

[3] Regulation (EU) No 910/2014 of the European Parliament and of the Council of 23 July 2014 on electronic identification and trust services for electronic transactions in the internal market and repealing Directive 1999/93/EC (OJ L 257/2014).

technology (Legal Geek 2018), merely revise the *method* of recording, transmitting, and processing of legal information, leaving the three remaining elements unchallenged. But this does not have to be the case.

New ICTs also open up possibilities to process legal information differently at the pre-publication stage. The most distinctive proposals in this regard suggest personalisation of legal information, i.e. bespoke tailoring of the subject-matter of publication (e.g. Casey and Niblett 2016; Ben-Shahar and Porat 2016; Porat and Strahilevitz 2014). This would mean creating of personalised rules and standards. For example, the current law can hold you liable if you cause damage to others by acting negligently. The negligence standard is now expressed in generic terms of "what would have a reasonable person done had this person been in your situation". If this legal information was personalised (e.g. by analysing big data sets containing information about yourself and by comparing them with information about past wrongdoers who were already held liable for breaching the negligence rule), it could be possible to create a more refined negligence standard for each individual. This processing of legal information could take place before another type of ICT disseminates it.

But there are other routes which one may take with regard to implementation of new ICTs in methods of publication of legal information. For instance, ICTs may be used to disseminate legal information more efficiently by lowering the importance of central databases (core nodes with legal information) and by letting the data run across the ICT-infrastructures more actively to reach its intended addressees. Methodically, ICTs can thus change the entire picture of publication models and we need to carefully consider at which stage of the existing model it could do so for the greatest benefit and with minimal harm. Alternatively, we should consider how to adapt the existing model to make full use of new ICTs. This, again, will need to be assessed against efficiency and sustainability of such model.

10.3.3 A Publisher

The digital challenge then inevitably raises questions regarding the publisher of legal information, because although the original author of the information could be an official authority, the publishers of such information are often private entities or black-boxed mechanisms. The pressing issue in this context is whether—once we start relying on new ICTs in respect of not only recording and transmitting but also processing of legal information—published information will remain authentic and not misinterpreted by such processes. The optimal model should thus carefully consider how such risks might be internalised, e.g. by not contracting out the publication responsibility to external entities which are driven by economic rather than efficiency or sustainability goals.

Another challenge is that multiplication of publishers (whoever they are) might lead to decentralised, divergent and, at the end of the day, incomprehensive publication practice. We know that this happened in the past and nothing suggests that the present era would be immune to such issues. The digital challenge, therefore, justifies that we revise the overall parameters of the publication model.

10.3.4 An Addressee

With new ICTs and their implementation in publication models, it is finally important to ask to whom legal information shall be delivered and thus also to whom it shall be published. Publication of legal information to the public at large (i.e. just making such information in principle publicly available) is no longer efficient and sustainable, as I tried to show in the introduction to this chapter. The problem is that the existing model is designed mainly to suit the needs of legal specialists. It is them to whom the law-making authority often speaks in the first instance, or at least that is how the model is designed. And although there are initiatives trying to facilitate free access to law around the globe, such as the Free Access to Law Movement, without sufficient specialist knowledge, it is hard for anyone to find a way through the piles of digitised online accessible legal documents.

The main addressees of these legal information systems thus are legal specialists and ordinary citizens usually get to know about existing rules via different channels such as TV, radio, school education, or by unpleasant encounters with law enforcing authorities. That, I think, is not an optimal situation since the addressees of legal information are not the same as addressees of publications of such information. There is an apparent and alarming discrepancy between those to whom the rules are published and those who shall obey them.

It seems to me, though, that the emerging ICTs now give us room to revise this model and to bring publication of legal norms closer to its addressees, e.g. by displaying relevant information in virtual locations where they should be applicable, by displaying them only when they should be applicable, or by displaying them only to those whom should apply them. Eventually, then, some legal information such as the organisational norms and standards (see Sect. 10.3.1) could be directly implemented in the ICTs which should comply with such standards.

10.4 A New Model for Digitalised Hyperlaw

The digital challenge compels us to re-think and, if necessary, also to re-design the current publication model in order to facilitate more efficient and sustainable flow of legal information that would reach more addressees. Modern technologies and digitalisation of legal sources also allow us to make these sources and rules truly public again. Looking back through my brief discussion of all four aspects of any publication model, it shall now be clear that the digital revolution and emergence of hyperlaw alongside new ICTs could spur such novel proposals. The discussion also revealed that any model should not be evaluated against its isolated aspects. By contrast, if a new model is to be efficient in publishing and communicating legal information and delivering it to its true addressees, i.e. to those who are expected to comply with given legal rules or standards, then our assessment of its design must jointly consider all four aspects: the subject-matter of publication, the method of publication, the publisher, and the addressee.

The basic idea behind the existing model is that people—as addresses of the relevant legal information—have the responsibility to know the laws and thus also have a duty to act accordingly, i.e. search for the information. If legal information will be delivered efficiently to its addressees in time, i.e. before they are expected to follow them, we could once again substantiate the idea that one's own ignoration of laws is not an excuse. But that would probably demand us to shift the burden of publication duties towards the state or public authorities, rather than to let the addressees to search actively for such information.

To keep such model not only efficient but also sustainable, it seems to me that one would need to decrease the total number of published legal information because a single person cannot read all the laws in her entire life and because most of these laws are simply not relevant to her. That, however, does not need to imply that legislation and case law, as we know it nowadays, needs to be reduced in their quantity. The problem could be dealt with at a different level. Let us remind ourselves for a moment of the example that I used in the introduction to this chapter. The image there was a personalised model of publication where you, as an addressee of a legal norm would be targeted and informed about such rule in time much in the same way as you now receive personalised advertisements. Note that this model does not require personalisation (and therefore multiplication) of legal information. On this account, legal systems could retain much of their features; only the rules and standards would be published and disseminated differently.

Moreover, it is possible to imagine that some standards and perhaps also some rules could be disseminated and implemented directly into new technologies. Such legal information would then become embedded in the informational hyperhistorical environment which we now create. As a result, a new model of publication of legal information could ensure that relevant legal information will be known and present in those locations where they should be applied. As a mature information society, we could simply eliminate some intermediaries (e.g. the intermediating legal information systems and their operators) and we could embed some regulations directly into the code of ICTs that process other legal information. This could, in effect, significantly lower the amount of information that human addressees need to process themselves.

Such envisaged model could be both efficient and sustainable, but without further research, it can bring additional ethical issues. On one hand, by returning legal information into the physical environment (by direct implementation), the model could bridge the digital divide between those who have access to ICTs and new technologies and those who do not have such access. On the other hand, such full shift to hyperlaw could make us fully reliant on ICTs and lead to, for instance, our loosing of many specialist skills and knowledge, or to diminishing responsibility of human agents (Yang et al. 2018: 11). One response to such ethical challenge could be to implement the new hyperlaw model of publication on top of the existing framework, rather than to see it as its replacement. This, however, could prove costly and unsustainable in the long run.

10.5 Conclusion

I hope to have persuaded the reader that the digital challenge gives us reasons to revise the existing models of publication of legal information. The current models are inefficient and unsustainable mainly because the existing publication methods are designed to serve legal professionals and not the ultimate addressees of legal information. This creates the unnecessary multiplication of intermediaries facilitating communication of legal information between the original law-making authority and the intended addressee. The problem thus is that publication models do not serve those who are expected to know the laws and comply with them. A new model should bridge this gap by bringing our attention back to the ultimate addressee of legal information.

Given the developments in ICTs, it is now possible to let the digital data reach its addressees and convey the relevant bits of legal information to them. In contrast with the existing models, it no longer seems necessary that people would search actively for legal information themselves. Instead, data conveying legal information may search for people—for the potential addressees. In addition, the developing ICTs make it possible, at least in theory, directly to implement some legal information by delivering it to an ICT that is expected to be compliant with the laws itself. The regulated technology could thus be a direct addressee of some legal information. Such solution could be referred to as an addressee-centric publication model. In contrast with existing research that promotes personalisation of the published legal content (personalised legal information), this proposal advocates personalisation of the entire publication model.

The purpose of this chapter was, however, only to outline the addressee-centric model as the potential next step in the evolution of publication models. The overall design of such (or any other new) publication model needs to be explored in future research should we want to take full advantage of the digital challenge, and should we want to make our publication model efficient and sustainable once again. In this regard, I tried to show what can be learned from the past, namely that publication of legal information is better thought of as part of a holistic model that contributes to "legal" maturing of our society. Besides, I showed that it could be beneficial to separate different aspects of any publication model (a subject-matter; a method; a publisher; an addressee) to see better how efficient and sustainable such model could be when we combine these aspects. A solid grounding for future work in this area can be found around international initiatives such as the Free Access to Law Movement, International Association for AI and Law, or Legal Geek, albeit they (so far) only focus on individual aspects of the publication model.

References

Bench-Capon, T., et al. 2012. A history of AI and law in 50 papers: 25 years of the international conference on AI and law. *Artificial Intelligence and Law* 20 (3): 215–319. https://doi.org/10.1007/s10506-012-9131-x.

Ben-Shahar, O., and A. Porat. 2016. Personalizing negligence law. *New York University Law Review* 91 (3): 627–688.

Biasiotti, M.A., and S. Faro. 2011. *From information to knowledge: Online access to legal information-methodologies, trends and perspectives.* Amsterdam: IOS Press.

Buringh, E., and J.L. Van Zanden. 2009. Charting the "Rise of the West": Manuscripts and printed books in Europe, a long-term perspective from the sixth through eighteenth centuries. *The Journal of Economic History* 69 (2): 409–445. https://doi.org/10.1017/S0022050709000837.

Casey, A.J., and A. Niblett. 2016. Self-driving laws. *University of Toronto Law Journal* 66 (4): 429–442. https://doi.org/10.3138/UTLJ.4006.

Charpin, D. 2010. *Writing, Law, and Kingship in Old Babylonian Mesopotamia.* Trans. J.M. Todd. Chicago: University of Chicago Press.

Cuccuru, P. 2017. Beyond bitcoin: An early overview on smart contracts. *International Journal of Law and Information Technology* 25 (3): 179–195. https://doi.org/10.1093/ijlit/eax003.

Eisenstein, E.L. 2012. *The printing revolution in early modern Europe.* 2nd ed. Cambridge: Cambridge University Press.

EUR-Lex. 2018. *Online statistics.* http://eur-lex.europa.eu/statistics/legislative-acts-statistics.html [https://perma.cc/WF8Q-QSN6].

Eurostat. 2018. *Population as of 1 January.* http://ec.europa.eu/eurostat/tgm/table.do?tab=table&init=1&language=en&pcode=tps00001&plugin=1 [https://perma.cc/NN7P-F4CS].

Floridi, L. 2012. Hyperhistory and the philosophy of information policies. *Philosophy & Technology* 25 (2): 129–131. https://doi.org/10.1007/s13347-012-0077-4.

Janeček, V. 2017. *Kritika právní odpovědnosti.* Prague: Wolters Kluwer.

———. 2018. Ownership of personal data in the internet of things. *Computer Law and Security Review* 34 (5): 1039–1052. https://doi.org/10.1016/j.clsr.2018.04.007.

Kramer, S.N. 1958. *History begins at Sumer.* London: Thames & Hudson.

Legal Geek. 2018. *Start-up map.* https://www.legalgeek.co/startup-map/ [https://perma.cc/HT35-C226].

Leith, P. 2016. The rise and fall of the legal expert system. *International Review of Law, Computers & Technology* 30 (3): 94–106. https://doi.org/10.1080/13600869.2016.1232465.

Lepri, B., et al. 2017. Fair, transparent, and accountable algorithmic decision-making processes. *Philosophy & Technology.* 31: 611–627. https://doi.org/10.1007/s13347-017-0279-x.

Porat, A., and L.J. Strahilevitz. 2014. Personalizing default rules and disclosure with big data. *Michigan Law Review* 112: 1417–1478.

Ruhl, J.B. 2018. Expanding the AI & Law Matrix. In: *Law 2050.* https://law2050.com/2018/02/19/expanding-the-ai-law-matrix/ [https://perma.cc/S32P-RT4F].

Wachter, S., B. Mittelstadt, and L. Floridi. 2017. Transparent, explainable, and accountable AI for robotics. *Science robotics* 2 (6): 1–2. https://doi.org/10.1126/scirobotics.aan6080.

Yang, G.-Z., et al. 2018. The grand challenges of science robotics. *Science robotics* 3 (14): 1–14. https://doi.org/10.1126/scirobotics.aar7650.

Chapter 11
From Bones to Bytes: A New Chapter in the History of Death

Carl Öhman

Abstract The informational remains of dead internet users are increasingly populating the web. To survey the most significant aspects of this phenomenon, the present chapter gives a historical background to the role death in society, and draws on current literature on the topic of digital afterlife. It is argued that the ongoing digital transformation of death takes place both on a microscopic, a macroscopic and a conceptual level.

11.1 Introduction

It is said that if you live by the digit you will also *die* by the digit (Floridi 2009, p. 52). And indeed, in a world where everything is electronic data, even *death itself* is going digital. In contrast to our biological remains, personal data do not decay after we die, but often live on through what has been characterized as a form of *digital afterlife*. To accommodate this phenomenon, a burgeoning body of tech companies has arisen, which I have previously referred to as the Digital Afterlife Industry (DAI) (Öhman and Floridi 2017). The service DeadSocial for example, promises to extend one's social media presence beyond the moment of one's biological death, and posts prewritten messages on designated occasions such as birthdays and graduations. More technologically advanced firms, such as Eterni.me and Replica, even allow users to create online chat-bots – artificial agents that, based on one's informational footprints, continue to impersonate the deceased after their death. Clearly, such technologies mark a major shift in how we relate to, and live with, the dead.

But digital Information and Communication Technologies (ICTs) also disrupt death on a societal level. In the next three decades, more than 2.7 billion people will pass away (United Nations 2017), and unlike previous generations, many of them will have lived parts of their lives online. In fact, nearly half of Earth's entire

C. Öhman (✉)
Oxford Internet Institute, Digital Ethics Lab, University of Oxford, Oxford, UK
e-mail: carl.ohman@oii.ox.ac.uk

© Springer Nature Switzerland AG 2019
C. Öhman, D. Watson (eds.), *The 2018 Yearbook of the Digital Ethics Lab*,
Digital Ethics Lab Yearbook, https://doi.org/10.1007/978-3-030-17152-0_11

population will be connected within 2018 already (Statista.com 2018), and they will undoubtedly leave a vast volume of data when passing. To manage these data in an ethical and just manner will be a difficult challenge, both for individuals, corporations, and society.

Despite its novelty, and perhaps due to its far-reaching consequences, online death is gaining increasing interest from academic audiences. It attracts researchers from fields as varied as law (Edwards and Harbinja 2013; Mccallig 2014), anthropology (Mitchell et al. 2016), and philosophy (Steinhart 2007; Stokes 2012). In this chapter I draw on this body of literature to survey the most significant aspects of the phenomenon, and provide some commentary on the limitations and potentials of said research. However, instead of structuring the presentation by academic discipline, I shall argue that online death is best understood through three different lenses: a macroscopic, a microscopic, and a conceptual one. Finally, I shall close the chapter with a set of recommendations for further research, and discuss its role in shaping sustainable regulation of the phenomenon. But before turning to the contemporary scholarship, some historical background is necessary.

11.2   Death in History

As remarked by the Spanish philosopher Miguel de Unamuno, "Stone was used for sepulchres before it was used for houses" (2005, p. 138). In other words, death has always occupied a central position in human culture. Despite this centrality, the conceptions and rituals surrounding it have shifted vividly throughout the centuries, often as a result of material and technological forces. Perhaps the most significant and oldest known example of such a shift is the emergence of the permanent settlements around 15,000 years ago. When abandoning the nomadic lifestyle, it became possible, even necessary, to keep the bodies of the dead closer to everyday life. In fact, many of the first settlers buried their dead, not only close to, but even *inside* their dwellings (Spellman 2014, p. 25). This new permanent proximity meant that the deceased maintained a greater social presence in the community which, as historian William Spellman puts it, helped shape the conviction that they "remained involved in community affairs" (2014, p. 25). For cultures such as the Natufian, this conviction was likely reinforced by the fact that the heads of the deceased were often also removed and kept as part of the descendants' home interior. While these practices may strike contemporary readers as bizarre, many of us still keep our ancestors close to everyday life, only we do it in the form of photo albums, and more recently perhaps – in cyber space.

With the emergence of the first cities, cultures, and organized religions, death became an increasingly abstract concept, imbued with cultural meaning and symbolism. Some scholars (Campbell 1972) have even hypothesized that the pursuit to give meaning to our mortality was the origin of mythmaking as such. And indeed, in all ancient societies we know of, death occupies a vital part of mythology, embodied by gods like Set and Hades and celebrated with monuments such as the great pyramids of Egypt and the Taj Mahal in India. It also constitutes the central theme in the world's arguably oldest surviving narrative – the Gilgamesh epos, which tells the tale of Mesopotamian half-god King Gilgamesh who, upon witnessing the gruesome death of his best friend Enkidu, sets out on a quest for eternal life. While getting close to succeeding in this venture, Gilgamesh eventually loses the key to eternity, and must face up to the inescapability of death. However, he does not return completely emptyhanded – he has gained the insight and wisdom to accept his, and every human's, place in the universe, as a finite and ultimately mortal entity.

The emphasis on acceptance of death and human limitation is a reoccurring theme throughout history, and especially in early philosophy. In Plato's *Phaedo*, for instance, Socrates famously argues that preparation for death and dying is the ultimate and true goal of those who practice philosophy "in the proper manner", and that fearing death is the worst of philosophical vices – the hubris of claiming to know what cannot be known. On a similar note, Epicurus (2005) taught his disciples that it was foolish to fear death since "when death is there, you are not", i.e. death cannot be bad for the living nor for the dead since none of them can possibly experience it. While there is probably little direct influence from the thinkers of antiquity, a similar attitude can also be found in early medieval Europe. As explained by Phillipe Ariès in his canonical book *Western attitudes toward death from the middle ages to the present* (1974), dying was then considered a rather normal element of everyday life, not as a disruption or a break, but as part of one's inevitable fate as planned by God. As an event, dying even followed a strict religious protocol focused on "understanding", and acceptance of one's fate. After this ritual, the deceased were more or less depersonalized, submerged within the collective of the Church – buried in anonymous graves.

Following the attitudes of early medieval Europe (what Ariés refers to as *Tamed Death*) he identifies three additional eras in the European attitudes to death: In the late middle ages, it is increasingly seen as *One's Own Death*, emphasizing the final judgement, and thereby individual biography, of the deceased. This changed again during the renaissance and romantic eras, when focus turned increasingly toward the death of others, what Ariés calls *Thy Death*. In the era of Thy Death (renaissance and romanticism), the purely Christian tradition was accompanied with more secular "rational" influences, with an increasing focus on the body and not just the soul. Hence, graves again became more personalised (after having long been neglected) and the custom of visiting individual tombstones started to emerge. The final era is that of modernity, which Ariés refers to as *Forbidden Death*. In modernity death becomes something to be hidden away. It should ideally take place at a hospital under the supervision of an expert doctor. Likewise, grief should ideally be accommodated by another "expert": the funeral director. However, despite making a compelling historical case, Ariés offers little insight into the forces involved in shaping the rituals.

11.2.1 Death as an Object of Study

In contrast to Ariés, sociologists have long been concerned with inquiries about the forces that shape our conceptions of death. Among the first to stress the societal nature of death in general, and mortuary rituals in particular was French sociologist Robert Hertz (1907). Hertz viewed mortuary rituals as a way for collective consciousness to re-identify the deceased, and in a sense, reintegrate them into the community – only now as *dead* members. Doing so, he thus stressed society's manifestation in individual and private events. Clearly, the pioneering (and arguably subject defining) work on suicide by Hertz' contemporary and colleague Emile Durkheim owes much to this approach. Durkheim argued that variation in suicide rates among different social groups could be explained by so-called *social facts*. War, for instance, has a significantly negative impact on suicide rates, as citizens normally become more integrated when faced with a common threat, and are allocated a social function for the group. In other words, the individual demise may be a personal tragedy, and is surely experienced by the bereaved as a unique and private matter, but this does not take away from death its social and cultural nature as a macroscopic phenomenon.

Since the days of Hertz and Durkheim, sociologists have kept on researching the social nature of death and the dead: Michel Kearl (1989, p. 111) has described the social function of death as a "double edged sword", that brings disruption to society, but also has the power to unite, reorganize and renegotiate power relations. Meanwhile, economic sociologist Viviana Zelizer (1978) famously traced the emergence of the American life insurance industry, investigating how the values, and rituals surrounding death relate to monetary value. Finally, a more modern example (plenty of important scholars are left out this brief presentation), Tony Walter (2015) argues that the social presence of the dead in part depends on the kind of ICTs available for that society, which – considering the possibilities realised by the digital revolution – means that we stand before a major shift in this presence. In sum, sociological studies of death are both numerous and diverse, and approach the topic both as a social, economic and technological event.

While the shifts described in this review took centuries (if not millennia) to take shape, current rituals surrounding death seem to evolve faster than what is possible for research can keep up with. This makes it difficult to pin down exactly how death is currently changing, and what direction we are heading towards. Nevertheless, there is reason to believe that the dead are again becoming increasingly present in our lives. Much like the first permanent settlers, we too are beginning to share our daily (virtual) living spaces with the dead. While the Natufians kept the skulls of the deceased in their houses, we keep their faces in our social networks. Undoubtedly, this new "proximity" to the digital remains of the dead, will have far reaching consequences. These consequences can be traced both at an individual (i.e. microscopic) and a societal (i.e. macroscopic) level. But, as we shall see, the technological leap also compels us to reconsider difficult questions about the nature of death as such, i.e. a conceptual level. In what remains of this chapter, I shall use these three levels as lenses to discuss the direction of what is perhaps the beginning of a new era in western (and indeed global) attitudes to death.

11.3   Online Death on a Micro-level

Let us begin by considering what we know about the transformation of death on a microscopic level. That is, how it is experienced by the individual herself and how it affects her inter-human relationships. This topic is naturally very multifaceted, but to simplify the presentation, I will use the chronological framework suggested by Gotved (2014). According to Gotved, the online death phenomenon can be understood as a timeline, consisting of three stages: Ante- (before); Peri- (during); and Post- (after) mortem. As we shall see, each stage of an individual's death is undergoing its own transformation.

11.3.1   Ante-mortem

As for the ante-mortem phase, digital technologies have most saliently increased the individual's ability to anticipate, plan, and dictate her death and afterlife. Last wishes, funeral arrangements and general preparations that would previously have required a fellow human trustee, can now be automated and executed remotely. There are a multitude of tech start-ups, so called Information Management Services (Öhman and Floridi 2017), devoted to facilitating such tasks. Many of which specialise in management of digital assets, which has become one of the largest posthumous issues to deal with for the digital human. A report released by the online security company McAfee (Smith 2011) showed that people put an average value of $37 000 ($55 000 for Americans) on their digital assets – a large and often confusing estate to manage. For this purpose, not only online services, but also self-help books such as Romano and Carrol's *Your Digital Afterlives* (2011) have emerged. However, apart from the fact that it is rarely considered by those to whom it lacks immediate relevance (Grimm and Chiasson 2014), little is still known about how digital technology affects individual *attitudes* toward death. For instance, does the ability to plan and prepare for death make us more, or less afraid of it? Is the idea of a digital legacy comforting or frightening? Are we about to exit the era of *Hidden Death?*

In one sense, it may be argued that Death is both becoming more *and* less private. In modern society (to Ariés *hidden death*) death has been removed from the family and one's home, and moved into institutions and hospitals. Instead of a gathered family, she who dies is most commonly surrounded by "expert" doctors at the final moment. Digital technologies enable one to go even one step further. For instance, a chat-bot service developed by start-up company Lifefolder, Emely, uses natural language processing to help users prepare and set everything in order before their death. "A conversation with Emily mirrors a conversation you would have with a trained nurse" Lifefolder explains (Kamps 2017). In one sense, such technologies make death even more hidden from human community than before, since it renders any human contact obsolete. But on other levels, such as the peri-mortem phase, death is instead becoming significantly *less* private.

11.3.2 Peri-mortem

In a purely biological sense, the Peri-mortem phase of death is (naturally) a very short event. Yet, socially, this is not necessarily the case. The Peri-morem phase is undergoing transformations with regards both to digitisation of traditional rituals, and to entirely new aspects to death. One example of the former is online funeral services, which began to emerge as early as the late 1990s (Sofka 1997). These services include both online bookings and tailoring of physical funerals, and remote video/virtual memorials for families who live far apart, making the demarcation between the virtual and the "real" increasingly blurred. While the funeral home, and the funeral director, used to play an important role in the Peri-mortem phase, the internet allows for an almost completely depersonalised process, without any encounter with fellow humans. Another example, also highlighting the increasing merge between the virtual and the "real", is the emergence of QR codes on tomb-stones (Gotved and Bjerager 2014), which, as put by Gotved and Bjerager (2014, p. 2) makes the tombstone "at once physical and digital, underhandedly putting presumably private content within public reach". Digital technology it seems, makes death *both* more private and public.

In addition to the migration and integration of traditional rituals into the digital realm, the internet also brings into existence entirely new aspects to the very event of dying. While death was always in part also a social event (Hertz 1907), online platforms add a new dimension to the rites of passage and renegotiation of identity that follow from death. For example, Keegan and Brubaker (2015) map out the process of turning "is" into "was" following the demise of notable persons on Wikipedia. According to their findings, the event of death sparks a large increase in revisions (i.e. renegotiations) of a person's biography, indicating the presence of a new kind of (online) rite of passage. To this, one may add a growing body of start-ups offering so called digital obituaries also for private consumers. In sum, death is increasingly an event taking place also in cyber space.

11.3.3 Post-mortem

In contrast to the physical body, digital human remains do not decay, and may thereby maintain a kind of (para-)social (Scherlock, 2013) relationship to the living. Many researchers have approached this relationship through the discourse of *continuing bonds* (DeGroot 2012; Getty et al. 2011; Kasket 2012) which originates from grief studies predating the breakthrough of the internet (Klass et al. 1996). It views grief, not so much as the reaction to an abrupt end, but as a process in which the relationship undergoes a metamorphosis, and becomes something new. As Kern et al. (2013, p. 2) notes, this is particularly salient on social media, which is increasingly becoming "a place to honor, memorialize, and *engage in dialogs* [my emphasis] with the deceased". "We do it to keep him alive", as one of Bell et al.'s (2015) interviewees puts it. Online grief has quickly become established as a customary social practice, one which appears to be disseminating exponentially too (Egnoto et al. 2014). In fact, death often means a large increase in social activity on a social media profile during the weeks following an individual's death (Castro and Gonzalez 2012). So, the dead are, in other words, becoming present in everyday life.

But online grief does not only take place on conventional social media platforms. There is a plethora of enterprises and platforms – so called online memorial sites – specifically devoted to accommodating it. To name but one representative example, the service imorial.com allows users to build a page onto which bereaved friends and family members may upload digital content such as videos, photos and text messages. Users may also form groups and send invitations to new users. However, it remains unclear whether this kind of service helps users get closure or if they in fact leads to prolonged grief. Many scholars are positive about the emergence of new digital afterlife services, and even complain that social media companies should put more effort into accommodating users' grief (Bell et al. 2015; Egnoto et al. 2014). Others however, warn that online memorials may have the opposite effect from what is promised. One example being Mitchell et al. (2014, p. 28) who, in studying specifically memorial sites devoted to deceased children argue that:

> [...] under the guise of addressing or even treating parental grief, on-line memorials may do more than simply accommodate that grief; they may perpetuate it. By enabling the deceased to persist, parenting to continue, and grief to be continually communicated, acknowledged and legitimated within a community of bereaved parents and a wider public, the Web affords an on-going grief that is unhinged partially from longstanding ideas of 'closure', privacy, and a separation of the living and the dead.

The question is however more nuanced than what is afforded by "the web". What matters is instead the particular design of individual services (and as we shall see later, the economic systems within which they operate). The design of memorial platforms are known to shape how users mourn their dead (Roberts 2012). As Acker and Brubaker (2014, p. 7) notes "platforms are never neutral tools because they privilege certain types of use with particular ends (e.g., commercial viability, vendor lock-in, or enrolling new users)". If these ends get priority over the wellbeing of the users, it may harm both the dignity of the deceased and the bereaved. For instance, McEwen and Scheaffer (2013, p. 1) describe some services as "environments of competition among mourners", which certainly gives one reason to worry. Nevertheless, despite their seemingly large influence on the way users interact with the online dead, little is still known about these services (Gotved 2014, p. 114), and how they negotiate between commercial- and other interests in the design process. This is arguably one of the most severe lacunas of current research.

To summarise, the individual, micro-level experience of death is undergoing a rapid transformation, both when it comes to planning, dying and remembering (interacting with) the dead. This transformation is largely accommodated by social media platforms, but also by specialised digital afterlife services. While there is a rather abundant knowledge about how people use these services, there is yet little research on the processes through which they are shaped. However, any observations or research into the micro-level phenomena, can only be properly understood when putting them in relation to the societal context.

11.4 Online Death on a Macro-level

Much like the micro-level, the macroscopic role of the dead in society is both multifaceted and difficult to pin down. However, at least some of the most salient aspects can be unpacked as a composition of a social, an economic and a political dimension. In other words, questions that arise at this level are social, insofar as the dead to an increasing extent remain interactive and present in society through their informational remains. They are economic, insofar as this data inevitably has to be stored and managed somewhere – and hence funded (by *someone*). They are political, insofar as the dead become stakeholders in political negotiations, and insofar as the control of their data in the end is also a control over history as such. While each of these dimensions comprises different phenomena, they do inevitably also tie into one another, and together form a larger development.

11.4.1 The Social Dimension

The dead have always been socially present one way or another, mediated through songs, books, recordings, laws, etc. As touched upon earlier in this chapter, their presence subsequently depends on the ICTs available to that society (Walter 2015). However, having one's personal information survive the demise of one's biological body has traditionally been a right granted only to a very limited group of people such as religious or political leaders. Today, this right is instead radically democratised – in information society, everyone is becoming a celebrity within their own network. Quantitatively speaking, we probably have more information recorded about any of the two billion Facebook users than we do about Jesus Christ or the Buddha (Siddhartha Gautama). While naturally, this does not guarantee every Facebook user an afterlife like that of Christ or the Buddha, it does indeed change at least the societal structures through which such an afterlife is made possible. Indeed, we are, as Jacobsen (2017) phrases it, about to enter *Postmortal society*.

The fact that we share cyberspace not only with other living users but also with past generations, can be seen in light of what Luciano Floridi (2014) has called "the fourth revolution" of our collective self-understanding. The fourth revolution follows upon the Copernican (we are not the centre of the universe), the Darwinian (we are not the centre of life) and the Freudian (we are not the centre of our minds), and denotes the fact that we are slowly acknowledging that we in fact share this world with other types of agents, artificial as well as follow biological ones. Most relevantly to this chapter, we also share this world with past and future generations, in the sense that what we do matters for all kinds of existences, for those who live, those who have lived, and for those who will. For some, this paradigm is interpreted as a form of "re-enchantment" of the world. Scherlock (2013, p. 173) for instance points out that the vast majority of Internet users have very limited understanding of how the technology behind it works, which opens up for "mythical interpretation like the notion of the Internet as digital heaven". In other words, the fourth revolution is not merely a philosophical concept, but a rather real experience among internet users.

While the information revolution has enhanced the ability of the dead to take place in our social institutions, technology does not singlehandedly determine how (or even *if*) this ability is realised. As I have pointed out elsewhere (Öhman and Floridi 2017), this is not simply a question of the affordances of certain technologies, but rather about their role in systems of human practice, and most importantly – their role in our *economic* systems.

11.4.2   The Economic Dimension

Any rigorous analysis of technology use must consider the economic systems in which said technology operates. The same is true for the ICTs mediating our informational remains. Subsequently, informational immortality often includes a kind of economic immortality too, at least when it comes to celebrity. According to Forbes (2015) for example, an artist like Michael Jackson has already "made" more than $140 million since his death in 2009 (from record sales, merchandise and even *hologram performances*). Similarly, John Lennon has posthumously made $12 million and Albert Einstein $11 million. So indeed, depending on what remains of their artistic or scientific contributions, celebrities do remain involved in economic affairs after their death. But – as touched upon in the previous section – not only does digital ICTs simply change the way celebrities are mourned and made present in society – everyone can now achieve a celebrity-like status within their own networks. Subsequently, everybody's afterlife can also be monetised.

Despite this development, little attention has hitherto been given to the economic aspects of online death. With exception for my own work (Öhman and Floridi 2017), and Aceti's (2015) somewhat related analysis of neoliberalism and the depiction of death, only two recent studies (as of 2018) discuss the economics of digital afterlife explicitly: Karppi (2013) and Meese et al. (2015). Karppi argues that social media sites are incentivised to keep dead users' information within their data economy rather than deleting it. Taking Facebook as his primary example, he further claims that in the "noopolitics" (2013, p. 14) of the site, the dead become "nodes that open up to other nodes and other agencies" in a networked economy. This is argued to be the reason behind Facebook's decision to memorialise profiles rather than deleting them (2013, p. 14). On a similar note, Meese et al. (2015, p. 416) (who do not engage in any explicit economic analysis) speak of an increasingly commercialised discourse of death in social media, arguing that this causes a "commercial and social push for the preservation of posthumous personhood".

While Karppi and Meese et al. have made important contributions, they tend to overlook one crucial aspect regarding the economics of online death, namely the costs surrounding the management and storage of digital remains. Within the next three decades, more than 2.5 billion people will die (United Nations 2017). Many of these will be internet users, and will subsequently leave behind a vast volume of data when passing. On Facebook for instance, the number of "dead" profiles has been estimated to reach a minimum of 1.4 billion by 2100 if Facebook ceases to grow, and with continued growth this number would be close to 5 billion (Öhman and Watson 2019). Elsewhere (Öhman and Floridi 2017), I have described digital remains as a form of capital, and argued that this quality may have far-reaching, and potentially harmful, consequences, especially since capital requires human labour to remain productive. That is, a growing volume of digital remains necessitates an increase in posthumous interaction online. For example, in 70 years, how will Facebook manage to fund the storage costs of two billion *dead* profiles? By using them to attract more living users? Undoubtedly, making the storage of 2 billion dead profiles financially viable requires a lot of labour. The next question to ask is: what are the political and macro-ethical implications of such a development?

11.4.3   The Political Dimension

Much like celebrities may acquire a kind of economic afterlife through their recorded information, so politicians can remain active in political systems. One needs look no further than the cult surrounding Vladimir Lenin, or Kim Il Sung's necrocratic leadership in North Korea, both of which exemplify how prolonged personal power can only be upheld by technological means (embalmment for Lenin, and written laws for Il Sung). Similarly, online death is also a matter of power, of control over narratives, over data, and ultimately – over history. An episode in George Orwell's novel *1984* (1949, p. 313), illustrates this well. While interrogated by Party member O'Brian, the protagonist Winston Smith argues that there exists an objectively true past. In response to this claim, O'Brian replies "Then where does the past exist, if at all?". "In records. It is written down." Winston explains. "In records. And- ?". "In the mind. In human memories." Winston adds. O'Brian's final argument concludes the episode: "In memory. Very well, then. We, the Party, control all records, and we control all memories. Then we control the past, do we not?". While one may disagree with O'Brian's metaphysical standpoint, it is hard to argue with the fact that informational archives set the frames within which history takes place.

Luckily, our current society has made sure to distribute historical data over a multitude of states and agencies. At least, there is no evil party that controls the records and human memories of all people. Yet, in the digital age, almost all our memories and records are instead stored and controlled by private technology empires. In today's mature information societies, almost everything is recorded for the sake of commerce. In difference to perhaps every other society to date, as described by Floridi (2014), the question is no longer about what information to record – but what to *delete*. That is, what information is no longer of interest? What is superfluous? As these questions become increasingly central to society, it is upon the academic community to add: *whose* interest? And superfluous for *what* purposes?

The answers to such questions are yet to take shape but undoubtedly depend on economic analysis. This point is stressed by Victor Mayer-Schönberger in his book *Delete* (2009); Today, he argues, it is simply cheaper to keep on storing data than to bother deleting it. Yet, it only holds because most people who have ever used the internet hitherto are still alive, and this is about to change. Will it still be cheaper to keep storing and never deleting content in say 40 years? And if not, will the hundreds of millions of African and Asian profiles, presumably visited by users with relatively lesser consumer power, be as valuable to keep as the European and American ones? If the management of digital remains is driven by capitalist logic, i.e. commercial interests, will this mean that those who fall outside of such interests also fall out of history? – Shall their data traces be erased? Such questions stand unanswered today. However, even systems seemingly devoid of capitalist logic are no guarantee for equality. For instance, Keegan and Brubaker (2015) discuss the politics of editing biographical Wikipedia articles after a person's death is a matter of *who* gets the right to edit *whose* story. Some of the findings indicate a significantly lesser interest devoted to people from Global south and female biographies. Furthermore, the lack of interest for Global South is also an urgent lacuna in academic literature as pointed out by Graham et al. (2013). Clearly, more research is needed here.

To sum up, the macroscopic level is about power. Following the fourth revolution in human self-understanding, digital technologies compel us to reconsider non-living persons as actors and stakeholders in our social, economic, and political institutions. They remain in our online communities and social networks, which in turn opens up for continued monetisation and commercialisation of their data. As a result of which, they must also be considered a *political* actor, a stakeholder in the political decisions we make as a society. Inevitably, such development requires a foundational conceptual analysis.

11.5   Online Death on a Conceptual Level

Online death is a complex and philosophically rich topic that cuts through many areas of study, such as philosophy of mind, personal identity, and ethics. Despite its foundational role in any proper analysis of online death, the conceptual level of analysis remains sparsely explored, at least among philosophers. For this reason, I will first briefly address some philosophical discussions, and thereafter turn to discuss how conceptual analysis can be practically implemented.

While death has occupied philosophers as long as there has been philosophy, it would be safe to say that technology has never really been part of the discussion.

For example, in *The Oxford Handbook of Philosophy of Death* (Bradley et al. 2012), the role of technology is barely even touched upon. However, online technologies now compel philosophers to reconsider questions of posthumous personhood, agency, and ethics anew. For instance, the fast progress of artificial intelligence and machine learning has led to some rather imaginative predictions such as so called "mind uploading" (Bell and Gray 2001) – installing one's consciousness on a computer – which in turn brings a form of digital immortality (Steinhart et al. 2015; Swan and Howard 2012). Steinhart (2007) even predicts that conscious "digital ghosts" will soon be inhabiting the web by the millions. However, many philosophers still consider such prospects to be at best a stimulating fantasy, and at worst a fundamental misinterpretation of what minds are and what computers are capable of (Hauskeller 2012). Regardless of which, the new forms of online posthumous existence (conscious or not) raise some difficult problems. Arguably, the most central questions are those regarding ethics, and more specifically posthumous harm: if the dead are, as argued in this chapter, increasingly involved in our social, economic, and political institutions, under what conditions should they participate – what, if any, are the rights of the online dead?

The question of posthumous harm has a long tradition within philosophy, and is perhaps most famously discussed by Epicurus (see Sect. 11.2). Since then, a variety of arguments have been raised both in opposition to his views (Feinberg 1974; Nagel 1970; Pitcher 1984), as well as in their favour (Taylor 2005). However, few philosophers, except Patrick Stokes (2015), have fully acknowledged how ICTs tie into the debate. Discussing the ethical status of digital remains, more specifically social media profiles, Stokes asks whether deletion of such content can be harmful to the dead. His response is non-epicurean, i.e. death and what happens to you after it *can* be bad for you. The argument takes its point of departure in Floridi's (2011) informational interpretation of personal identity, which holds that an individual should be understood as an informational, self-aware entity. From this perspective, it makes sense to say that while death deprives an individual of their consciousness and bodily functions, the *person,* seen as a body of information, survives the death of the biological body. According to Stokes, this means that digital remains, just like physical bodies, are part of the personal identity of the deceased. Furthermore, such remains are (in accordance with Floridian ethics [2013]) argued to hold intrinsic value and a right to ethical respect in themselves, meaning that their deletion should be regarded as a form of "second death".

11.5.1 Practical Implementations

The nature of regulatory responses to online death depends on how the problem is conceptualised. I have discussed this at some length elsewhere (Öhman and Floridi 2018). While I do not subscribe to Stokes' notion of deletion of one's digital remains as a "second death", I do agree with his interpretation of posthumous personhood, and the idea of digital remains as a form of informational corpse. Much like corpses, digital remains cannot be "killed", but this does not mean they are immune to harm. Thus I propose that regulators should seek inspiration from the ethical frameworks that apply to *organic* human remains, such as archaeological museums. As many collections are now digitalised and made available online (Alberti and Hallam 2013), the ethical concerns of archaeology appear to be increasingly merging with those of the DAI. Luckily, the former already has regulatory frameworks in place such as the ICOM code of ethics (1986).

One of the central ethical concepts in the ICOM code is *human dignity*. Translated to the realm of the DAI, human dignity requires that digital remains, (here seen as the informational corpse of the deceased), may not be used *solely* as a means to an end, such as profit, but regarded instead as an entity holding an *inherent* value. The ICOM Code specifies that "all aspects of commercial ventures" must be carried out with respect for "the intrinsic value of the original object". Adopting a similar regulative approach to the DAI would clarify the relationship between deceased individuals and the firms displaying their data in several ways. Despite being the sole owner of the data, and irrespective of the desires of those next to kin, firms would be obliged to abide by certain conventions, such as preventing hate speech, and abstaining from commercial exploitation of digital remains. While the archaeological approach gives some indication as to the direction of future research, there is still much to be done. What is the ontological status of digital human remains? What is the macro-ethics of past generations? How does dignity relate to deletion? Answering these questions is an important first step in shaping the systems through which the dead become present in society.

11.6   Concluding Comments

Be it in the shape of bones or bytes, the dead have always remained present one way or another in the life world of the living. Much like ancestral skulls kept the dead involved in the community affairs of the first permanent settlers, so do the digital remains of our dead keep them involved in our online communities. In this chapter I have argued that this shift has consequences both on an individual (micro), societal (macro) and conceptual level. To the individual, it manifests itself in an increasing ability to plan one's afterlife, and as a continued bond to deceased friends and family online. On a societal level, these continued bonds are increasingly interwoven with the networks of economics and politics, which in turn compel us to ask difficult conceptual questions about posthumous personhood and posthumous harm. These phenomena are undergoing a fast development as we speak. It is therefore of uttermost importance that we take control over it and make it our own, a task that will require careful but unorthodox conceptual engineering.

References

Aceti, L. 2015. Eternally present and eternally absent: The cultural politics of a thanatophobic Internet and its visual representations of artificial existences. *Mortality* 20 (4): 319–333.

Acker, A., and J.R. Brubaker. 2014. Death, memorialization, and social media: A platform perspective for personal archives. *Archiv* (77).

Alberti, S., and E. Hallam. 2013. *Medical museums: Past, present, future*. Retrieved from: https://www.researchgate.net/publication/273506763_Medical_Museums_Past_Present_Future.

Ariès, P. 1974. *Western attitudes toward death from the middle ages to the present*. Baltimore: Johns Hopkins University Press.

Bell, G., and J. Gray. 2001. Digital immortality. *Communications of the ACM* 44 (3): 28–30. https://doi.org/10.1145/365181.365182.

Bell, J., L. Bailey, and D. Kennedy. 2015. "We do it to keep him alive": Bereaved individuals' experiences of online suicide memorials and continuing bonds. *Mortality* 20 (4): 375–389.

Bradley, B., F. Feldman, and J. Johansson. 2012. *The Oxford handbook of philosophy of death*. Oxford/New York: Oxford University Press. https://doi.org/10.1093/oxfordhb/9780195388923.001.0001.

Brown, K.V. 2016. *We calculated the year dead people on Facebook could outnumber the living*. Fusion.net. Retrieved from: http://fusion.net/story/276237/the-number-of-dead-people-on-facebook-will-soon-outnumber-the-living/.

Campbell, J. 1972. Myths to live by http://s3.amazonaws.com/s3.edu20.org/files/202167/Joseph%20Campbell%20-%20Myths%20To%20Live%20By.pdf?AWSAccessKeyId=AKIAJL2YKQD4VUAFRMRQ&Expires=1522401992&Signature=UTPRF4x7VNS2K%2BPtTpnXWa4i4xs%3D.

Castro, L.A., and V.M. Gonzalez. 2012. Afterlife presence on Facebook: A preliminary examination of wall posts on the deceased's profiles. In *CONIELECOMP 2012, 22nd international conference on electrical communications and computers*, 355–360. IEEE. https://doi.org/10.1109/CONIELECOMP.2012.6189938.

DeGroot, J.M. 2012. Maintaining relational continuity with the deceased on Facebook. *OMEGA–Journal of Death and Dying* 65 (3): 195–212.

Edwards, L., and E. Harbinja. 2013. Protecting post-mortem privacy: Reconsidering the privacy interests of the deceased in a digital world. *Cardozo Arts & Entertainment Law Journal 32*: 83–129.

Egnoto, Michael J., Joseph M. Sirianni, Christopher R. Ortega, and Michael Stefanone. 2014. Death on the digital landscape: A preliminary investigation into the grief process and motivations behind participation in the online memoriam. *OMEGA - Journal of Death and Dying* 69 (3): 283–304.

Epicurus. 2005. *Letters and Sayings of Epicurus*. Trans. O. Makridis. New York: Barnes and Noble.

Feinberg, J. 1974. The rights of animals and unborn generations. In *Philosophy and environmental crisis*, 43–68.

Floridi, L. 2009. Hyperhistory and the philosophy of information policies. In *Onlife manifesto*, ed. L. Floridi, 51–64. Cham: Springer. https://doi.org/10.1007/978-3-319-04093-6.

———. 2011. The informational nature of personal identity. *Minds and Machines 21* (4): 549.

———. 2013. *The ethics of information*. Oxford: Oxford University Press.

———. 2014. *The 4th revolution: How the infosphere is reshaping human reality*. Oxford: Oxford University Press.

Forbes. 2015. *Top-earning dead celebrities 2015*. Forbes.com. Retrieved from: http://www.forbes.com/dead-celebrities/#524884ce6a51.

Getty, Emily, Jessica Cobb, Meryl Gabeler, et al. 2011. I said your name in an empty room: grieving and continuing bonds on Facebook. In *Proceedings of the SIGCHI conference on human factors in computing systems*, 997–1000. New York: ACM.

Gotved, S. 2014. Research review: Death online – Alive and kicking. *Thanatos 3* (1): 112–126.

Gotved, S., and K. Bjerager. 2014. Privacy with public access: digital memorials on QR codes. Paper presented at Internet Research 15: The 15th Annual Meeting of the Association of Internet Researchers. Daegu, Korea: AoIR. Retrieved from http://spir.aoir.org.

Graham, C., M. Gibbs, and L. Aceti. 2013. Introduction to the special issue on the death, afterlife, and immortality of bodies and data. *The Information Society 29* (3): 133–141. https://doi.org/1 0.1080/01972243.2013.777296.

Grimm, C., and S. Chiasson. 2014. *Survey on the fate of digital footprints after death.* Usec'14, February. https://cs.carleton.ca/sites/default/files/tr/TR-14-01.pdf.

Hauskeller, M. 2012. My brain, my mind, and I: Some philosophical assumptions of mind-uploading. *International Journal of Machine Consciousness 4* (1): 187–200. https://doi.org/10.1142/S1793843012400100.

Hertz, R. 1907. *Death and the right hand.* Aberdeen: Cohen and West.

International Council of Museums (ICOM). 1986. *Code of professional ethics.* http://archives.icom.museum/1986code_eng.pdf. Accessed 1 Aug 2016.

Jacobsen, M.H. 2017. *Postmortal society: Towards a sociology of immortality.* Abingdon/New York: Routledge, an imprint of the Taylor & Francis Group.

Kamps, H.J. 2017. *Introducing Emelie–The chatbot that talks about end of life.* Medium, 20 June. https://medium.com/life-folder/introducing-emily-the-chatbot-that-talks-about-death-97b390119cce.

Karppi, T. 2013. Death proof: On the biopolitics and Noopolitics of memorializing dead Facebook users. *Culture Machine* 14: 1–20.

Kasket, Elaine. 2012. Continuing bonds in the age of social networking: Facebook as a modern-day medium. *Bereavement Care* 31 (2): 62–69.

Kearl, M. 1989. *Endings: A sociology of death and dying.* Oxford: Oxford University Press.

Keegan B., and J. D. Brubaker. 2015. 'Is' to 'Was': Coordination and Commemoration in Posthumous Activity on Wikipedia Biographies. Proceedings of the 18th ACM Conference on Computer Supported Cooperative Work & Social Computing, March 14–18, 2015, Vancouver, BC, Canada. https://doi.org/10.1145/2675133.2675238.

Kern, R., A.E. Forman, and Gil-Egui. 2013. RIP: Remain in perpetuity. Facebook memorial pages. *Telematics and Informatics* 30 (1): 2–10.

Klass, D., S.L. Nickman, and P.R. Silverman. 1996. *Continuing bonds, new understandings of grief.* Washington, DC: Taylor & Francis.

Mayer-Schönberger, V. 2009. *Delete: The virtue of forgetting in the digital age.* Princeton: Princeton University Press.

Mccallig, D. 2014. Facebook after death: An evolving policy in a social network. *International Journal of Law and Information Technology* 22: 107–140. https://doi.org/10.1093/ijlit/eat012.

McEwen, R., and K. Scheaffer. 2013. Virtual mourning and memory construction on facebook: Here are the terms of use. *Proceedings of the ASIST Annual Meeting* 50: 1–10. https://doi.org/10.1002/meet.14505001086.

Meese, James, Bjorn Nansen, Tamara Kohn, Michael Arnold, and Martin Gibbs. 2015. Posthumous personhood and the affordances of digital media. *Mortality* 20 (4): 408–420.

Mitchell, Lisa M., Peter H. Stephenson, Susan Cadell, and Mary Ellen Macdonald. 2014. Death and grief on-line: Virtual memorialization and changing concepts of childhood death and parental bereavement on the internet. *Health Sociology Review* 21 (4): 413–431.

Mitchell, L.M., P.H. Stephenson, S. Cadell, M.E. Macdonald, L.M. Mitchell, P.H. Stephenson, … M.E. Macdonald. 2016. Parental bereavement on the Internet, *1242* (April). https://doi.org/10.5172/hesr.2012.21.4.413.

Nagel, T. 1970. Death. *Noûs 4* (1): 73–80.

Öhman, C., and L. Floridi. 2017. The political economy of death in the age of information: A critical approach to the digital afterlife industry. *Minds and Machines* 27: 639–662. https://doi.org/10.1007/s11023-017-9445-2.

————. 2018. An ethical framework for the digital afterlife industry. *Nature Human Behaviour* 2: 318–320. https://doi.org/10.1038/s41562-018-0335-2.

Öhman, C., and D. Watson. 2019. Are the dead taking over Facebook? A big data approach to the future of online death. *Big Data & Society*. https://doi.org/10.1177/2053951719842540.

Orwell, G. 1949. *1984*. Retrieved from: https://www.planetebook.com/freeebooks/1984.pdf.

Pitcher, G. 1984. The misfortunes of the dead. *American Philosophical Quarterly* 21 (2): 183–188.

Roberts, P. 2012. "2 people like this": Mourning according to format. *Bereavement Care* 31 (2): 55–61. https://doi.org/10.1080/02682621.2012.710492.

Sherlock, A. 2013. Larger than life: Digital resurrection and the re-enchantment of society. *The Information Society* 29 (3): 164–176. https://doi.org/10.1080/01972243.2013.77730.

Smith, J. 2011. McAfee reveals average Internet user has more than $37,000 in underprotected digital assets, September 27. http://www.mcafee.com/us/about/news/2011/q3/20110927-01.aspx.

Sofka, C.J. 1997. Social support "internetworks," caskets for sale, and more: Thanatology and the information superhighway. *Death Studies* 21 (6): 553–574. https://doi.org/10.1080/074811897201778.

Spellman, W.M. 2014. *A brief history of death*. Retrieved from https://ebookcentral.proquest.com.

Statista.com. 2018. *Worldwide internet user penetration from 2014 to 2021*. https://www.statista.com/statistics/325706/global-internet-user-penetration/.

Steinhart, E. 2007. Survival as a digital ghost. *Minds and Machines* 17 (3): 261–271. https://doi.org/10.1007/s11023-007-9068-0.

Steinhart, E.E., P. Stokes, E.E. Steinhart, W. Youyou, M. Kosinski, D. Stillwell, et al. 2015. Your digital afterlives: Computational theories of life after death. *Minds and Machines* 17 (4): 363–379. https://doi.org/10.5172/hesr.2012.21.4.413.

Stokes, P. 2012. Ghosts in the machine: Do the dead live on in Facebook? *Philosophy and Technology* 25 (3): 363–379. https://doi.org/10.1007/s13347-011-0050-7.

Stokes, Patrick. 2015. Deletion as second death: The moral status of digital remains. *Ethics and Information Technology* 17 (4): 237–248.

Swan, L.S., and J. Howard. 2012. Digital immortality: Self or 0010110? *International Journal of Machine Consciousness* 4 (1): 245–256. https://doi.org/10.1142/S1793843012400148.

Taylor, J.S. 2005. The myth of posthumous harm. *American Philosophical Quarterly* 42 (4): 311–322.

Unamuno, M. 2005. *Tragic sense of life*. Gutenberg Project. http://archive.org/stream/tragicsenseoflif14636gut/14636.txt.

United Nations, Department of Economic and Social Affairs, World Population Prospects: The 2017 Revision, custom data acquired via website. 2017. World population prospects: The 2017 revision, custom data acquired via website.

Walter, Tony. 2015. Communication media and the dead: From the stone age to Facebook. *Mortality* 20 (3): 215–232.

Wróbel, Maria, Patrycja Bronowicka-Adamska, and Anna Bentke. 2017. Hydrogen sulfide generation from L-cysteine in the human glioblastomaastrocytoma U-87 MG and neuroblastoma SHSY5Y cell lines. *Acta Biochimica Polonica* 64 (1).

Zelizer, V. 1978. Human values and the market: The case of life insurance and death in 19th-century America. *American Journal of Sociology 84* (3): 591–610.

Chapter 12
The Green and the Blue: Naïve Ideas to Improve Politics in a Mature Information Society

Luciano Floridi

12.1 Introduction

In this chapter, I present some ideas that I hope may help improve political thinking and practice in a mature information society.[1] The ambition is quintessentially philosophical: trying to understand and improve the world, to the extent that each of us can contribute, in this case with some intellectual work. That is all. It is not a little, I realize, but it is not much either. It is the usual paradox: how important is a vote, or, in this case, a conceptual contribution? As much as a grain of sand on the beach: one counts for nothing, two are still nothing, but millions of grains can make a significant difference, if only because, without them, the beach would not exist. This is the *relational value of* aggregation. The ambition is therefore philosophical, but also aggregative, because I hope that the ideas expressed in this chapter may be useful and find some follow-up.

The ideas presented are philosophical, but they want to avoid being too abstract, so as not to be ultimately inapplicable. However, they do not want to be overly applied either, because it is up to a government to discuss and transform ideas into specific political actions. The point is to find the right distance between politics as pure political science and politics as a practice of *policy*. For this reason, the correct term to describe the ideas in this chapter can be borrowed from medicine, where the most abstract theory of a Nobel laureate and the most applied practice of a family doctor are never dissociated: they are *translational* ideas. They have the objective of articulating a foundational reflection that can be translated into concrete strategic

---

[1] It should be clear contextually, but let me clarify that in this chapter I only refer to good ideas that can influence politics, not to any good ideas in general, for example scientific ideas.

L. Floridi (✉)
Oxford Internet Institute, Digital Ethics Lab, University of Oxford, Oxford, UK

The Alan Turing Institute, London, UK
e-mail: luciano.floridi@oii.ox.ac.uk

© Springer Nature Switzerland AG 2019   183
C. Öhman, D. Watson (eds.), *The 2018 Yearbook of the Digital Ethics Lab*,
Digital Ethics Lab Yearbook, https://doi.org/10.1007/978-3-030-17152-0_12

guidelines, for the realization of specific political, legislative, economic, organizational, and technical actions. It is not an original idea: good philosophy has tried to be translational at least since the time of Socrates. We only lacked the right word.

Offering ideas for improving politics is a political operation in itself. This is because today politics is emerging more and more as *a relational activity* (the central theme of this chapter), and it is typical of some of the relational phenomena to absorb also their negation. For example, lack of interaction is a form of interaction, as an omission; lack of communication is a form of communication, because silence also speaks volumes about who is silent, and about what they are silent; and lack of information is a form of information, because it has a communicative value, given that a question without an answer is always informative with respect to the need to know something, and to its lack of satisfaction. Politics belongs to this kind of relational phenomena. Not doing politics—for example abstention—still remains a political behaviour, at least insofar as it delegates political decisions to others. It is therefore an illusion to think that we can live in a society and not be political. Only solitude can be genuinely apolitical (not solipsism, which is only the state of *believing* that one is alone). Even with only two people in a desert island, like Robinson Crusoe and Friday, politics is already inevitable. For this reason, Aristotle was partly right: we are all political animals, because even the attempt not to be political remains a political act. But he was wrong in thinking that we are *voluntarily, continuously*, and *rightly so*. None of the three conditions is ever entirely taken for granted, and today all are unfulfilled, for the following reasons.

First, because, in every democracy that exists today, we are political even *involuntarily*, that is, against our explicit will, not only unconsciously. And this can generate irritation and conflict since we cannot escape politics even when we want to reject it because it has disappointed us and we do not like it.

Second, because, in a *mature* information society (a concept to which I will return later), we are never "always political", but more and more often we are political *intermittently*, when social attention is called to express its judgment. For this reason, the communication mechanisms of politics are almost indistinguishable from the communication mechanisms of marketing, especially in countries where comparative advertising is permitted ("this product is better than that one"). The medium pursues the same goal: to attract or renew and then to keep the attention of the people (be they clients or citizens) on a particular theme, be it a new product or a new political issue. If this happens often, the result is a constant renewal of the stimulus, which requires increasingly intense doses to have some effect. Marketing has its own pace, and so does politics. Nobody launches a new product casually if can dictate the timescales of innovation. One must allow at least 12 and possibly 24 months to pass, so that people become accustomed to a product and lose the memory of its novelty. A great example is provided by the scheduling of a new iPhone (see Fig. 12.1).

After a while the old model is taken for granted, and at that moment it becomes replaceable by the new model. This shifts the risk of obsolescence: it is not the product that becomes obsolete, but rather the users who are not up to date with the latest model. It is not the product that ages, because this is constantly renewed, but

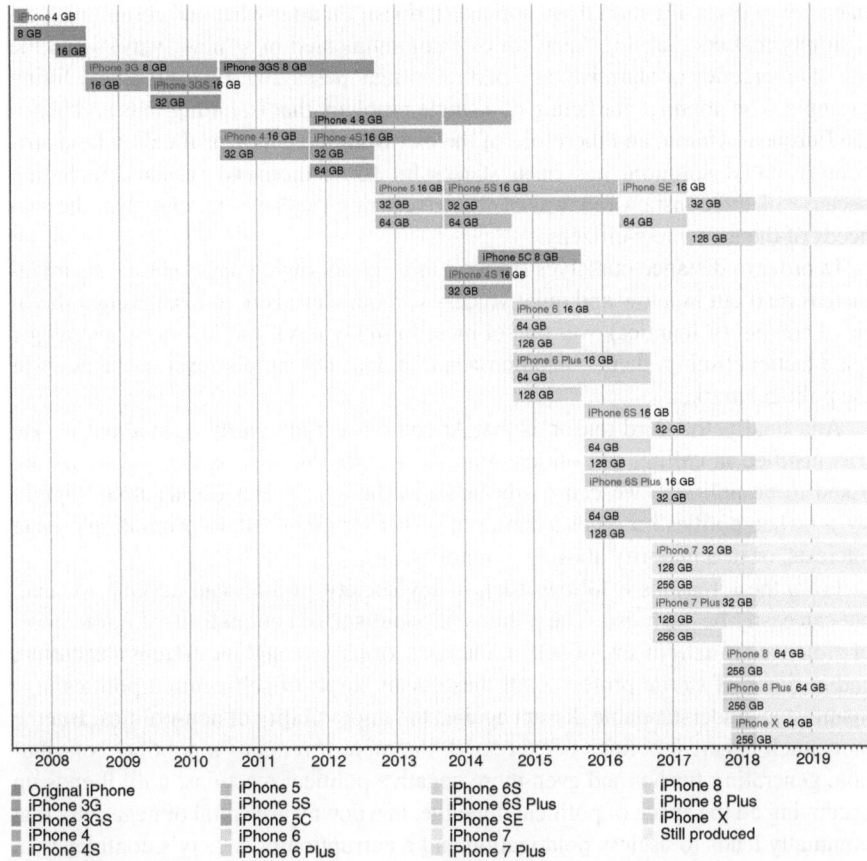

Fig. 12.1 Timeline of iPhone models. (Source: Wikipedia, https://en.wikipedia.org/wiki/Template:Timeline_of_iPhone_models and Apple Inc. (2004–2016). Press Release Library. April 6, 2016)

its users. Apple enjoys a position of leadership because it is able to control the timing of innovation, not suffer it—and vice versa, it can control the time of innovation because it is in a position of leadership. But those who have the competition snapping at their heels may not have the luxury of dictating the timing of innovation, be it a commercial or a social offer. This is often the case with politics, and today it seems the norm. If politics is the constant pursuit of populist consensus, each political actor will inevitably be tormented by his or her competition, and therefore devoid of control over their diary and agenda. Every move must be countered by another move, everyone chasing one another, no one driving. People do not become used to political solutions but only addicted to the communication that advertises them (or indeed that advertises new political *problems*). Not all current politics shares this asynchrony, but the fact remains that today we need the political call to take action; and yet we are also addicted to this call, requiring an ever-increasing intensity or diversification to be noticed (it doesn't matter what one communicates, it simply matters that it is "new" or else communicated in a "new" way)—and the use of emergency or alarmism as a part of normal messaging. Brexit is a very fitting example. The populist marketing of a single problem, that is, immigrants and hence the European Union, and therefore of the exit from the Union as the only (and stridently) stated solution, has been successful for a thousand reasons, including because of the constant renewal of the advertising message, unrelated to the real needs of the customers-citizens.

In order to distance ourselves from all this, the following pages should be imagined as read out in a low and quiet voice—without alarmism, re-evaluating a rhetoric of content (semantics) over that of mere form (syntax), and favouring a strategic not a tactical timing, that is, an approach that does not simply react to the news of the political market.

And finally, the third reason is that Aristotle was only partly right about us. He was justified in calling us political animals, but the problem is that, when we are asked to be political, we can easily be so in the *wrong way* (rather than "rightly so")—when politics becomes a matter of power serving itself, of promoting private interests, or of a majority abusing a minority.

From these reasons it follows that, in any society, politics can never be denied, but can easily be degraded. The politics of populism, of nationalism, of intolerance, of violence, of extremism, of selfish interests, of passive and indifferent abstention, and at times of sterile protest... all these many kinds of self-centred politics *also* manifest an understandable dissent against the impossibility of non-politics. But the more such negative politics is expressed, the more it remains a political contribution, generating further and even more negative political reactions, until it ends up occupying all the space of political dialogue, in a downward spiral of negativity that eventually leads to useless polarization and a corruption of society's confidence in its political abilities. Today, there is no lack of good policies to be realized, because there is a lot of intelligence around. What is missing is the right approach to remove the obstacles to implementation, because goodwill, while as abundant as intelligence, has become estranged from politics. By not doing politics, goodwill turns to self-destruction, because it leaves room for bad politics, which in turn negatively influences the exercise of goodwill. The frustration of reason joins the optimism of the heart[2] in regretting so many opportunities wasted, while the world is in such great need of them.

In the light of these problems, the political ideas expressed in this chapter are intended to be constructive, non-destructive, and *super partes*, not party-oriented or ideological. Not for anti-party reasons. As I argued above, anti-partitism and anti-politics now belong to the most widespread and sometimes "smart" partisan and political rhetoric. But because these ideas, to the extent that they can be useful, are offered to any political force that is interested in using them to govern better. In other words, the ideas presented here are *open source* and without constraints: adoptable and adaptable by anyone who thinks that they may have some value.

---

[2] The implicit reference to Gramsci is meant.

The title of this chapter takes up an idea, expressed in an article I wrote some time ago, on the need to unite green environmental policies (green economy and sharing economy) with blue digital policies (service economy), in favour of an economy of experience, that is, centred on the quality of relationships and processes, and not so much of consumption, that is, not so much centred on things and their properties. These are topics to which I shall return in the following pages. Here, I would like to explain in what sense the ideas offered would like to be *naïve*.

The ideas presented are naïve not in the sense that they are *void* of any "cunning of reason"[3] in the clever calculation of conveniences, or of opportunistic cynicism in the evaluation of power. But that they were intentionally *emptied* of it, in retrospect, and with disenchantment, but without disappointment. Think of the difference between a new coffee machine, which is void because it has never made any coffee, and a used and empty one, which contains no coffee because it has been emptied. Coffee tastes better when made in the second one, the one that has been emptied, not in the first, which is still void. In other words, the patina of reflection improves itself. This is why historical memory has enormous value, as a reminder of a presence of meaning, which requires a mental life to be appreciated, and not as a mere recording of facts, for which a digital system is sufficient.

This emptying—or "naïve-fication" to use a neologism—has been pursued in this chapter to give space to social altruism; to the intergenerational pact; to care for the world; to the sense of common homeland; to civil and ecological liability; to the political vocation as a service towards institutions, the State, and the *res publica*; to a cosmopolitan and environmentalist vision of *the human project*, understood as a society and life that we would like to see realized in the world; and finally the possibility of talking about good and bad politics. These *relations* are all qualified by many values, as we shall see later. Today it takes courage to use these expressions, because political ingenuity is seen as nonsense, for incompetent beginners, or as crafty cunning, for cynical politicians. Many deride it, or suspect it to be mere rhetoric, behind which other meanings, ambitions, messages, or manoeuvres can be hidden, to be deciphered according to the refined art of the most advanced political gaming. These many can stop reading this chapter. It is not written for them, because it means only what it shows and does not intend to show anything but what it says,[4] with the simplicity that should qualify the most serious and mature politics. Or as Paul of Tarsus says in his *Letter to Titus*: "Everything is pure for the pure [*Omnia munda mundis*]; but nothing is pure for the polluted and infidels; their mind and their conscience are contaminated (1:15)". The contaminated should take no offence, but they will not understand it.

By adopting this "naïve" approach, this chapter does not disregard Machiavelli or Hobbes, Locke, Rousseau, Kant, Mill, or Rawls. Yet in the end it is based, laically, on the most forward-looking strategy contained in *Matthew 18:3* "if you do not become like children you will never enter". *Ingenuousness* (naivety) is the point from which we start and to which we must return as *ingenuity* after the enriching

---

[3] The implicit reference to Hegel is meant.

[4] The implicit reference to Wittgenstein is meant.

path of reflection. It is sometimes the highest degree of sophistication to which we can aspire. Ithaca is a good analogy. And if this "forward" return to naivety (not backwards regress) perhaps cannot save the soul, maybe it can save politics. For this reason, a more adequate title for this chapter could be, in a less arrogant and ambitious way, "ideas that *would* like to be naïve".

Europe needs good ideas for a political government strategy that values and promotes its potential at best, not so much as a post-industrial society, but as a mature information society. The Union is emerging from a long period of crisis, at least as regards the economy, if not also the social aspects (especially in terms of the fracture of the social pact, even intergenerational, reduction and impoverishment of the middle class, less social mobility, and polarisation of opportunities that are not fairly equal), political (especially in terms of crisis of trust in institutions, populism, and personalization of politics), and cultural (national identity, immigration, role of Europe in a globalised world). In this delicate phase of recovery, the point is not being original at all costs, or imitating the US or China or other political realities, but recognizing and taking full advantage of the specific strengths of the many Europes that the EU contains, while reducing their weaknesses, and above all identifying the obstacles that do not allow these two operations. In light of this strategy, the wish is that the following naïve ideas, offered to improve policy, will be of some help.

To simplify the reading, the chapter is divided into nine sections: this introduction, now coming to an end; the conceptual framework that gives meaning and explains the following sections; the introduction of the idea of a mature information society, which today requires a new human project; a brief explanation of the idea of a human project and what it can mean today; the introduction of the idea of infra-ethics, understood as the structure that facilitates good moral action and a necessary factor to realize the human project; the presentation of some ideas to improve politics in a mature information society, together with the definition of some key concepts and principles that would underlie a good policy; the conclusion; and the acknowledgements.

I avoided as much as possible technical expressions and bibliographic references. They do not serve but hinder the development of ideas and the flow of reasoning. Philosophy is *conceptual design*. At its best, it analyses fundamental problems—that is, those richer in consequences (like the first dominos in a chain)— and articulate, in a factually correct and logically cogent way, solutions that are always open to sensible, informed, and urbane discussion, because the problems in philosophy are intrinsically open. Scholarly and rhetorical trappings unnecessarily burden it, hiding its rational and functional structure,[5] and I have therefore tried to avoid them.

---

[5] The implicit reference to the Bauhaus is meant.

12.2   The Idea of a Transition from Things to Relationships

Our way of thinking—especially in economics, law, politics, and sociology—is still dominated by a profound and implicit philosophy of an Aristotelian and Newtonian nature, and it is now obsolete. In order to label it, we may conveniently refer to this "philosophy behind philosophy"—this conceptual paradigm that we do not question when we do philosophy—as our *Ur-philosophy*.

The Aristotelian and Newtonian Ur-philosophy has worked well in the past: our way of thinking is still unknowingly formatted by it precisely because of its great success. Let's see it briefly, to understand why it would be a mistake to continue to apply it (perhaps by adapting it), to try to extract from it the right answers to the new political questions posed by the information society.

An Aristotelian Ur-philosophy conceives of society as lego-like in structure. There are many units of bricks that connect to other units of bricks, from the bottom to the top, to create complex structures, interacting with each other. Bricks or atomic (i.e., not further divisible) entities are *natural* or *legal* persons. And their various combinations are the couple, the family, a generation, a social class, an ethnic group, an industrial sector, an administration, a political party, and so on. The properties (what qualifies the bricks for what they are, for example "age", or "is a company") and behaviours (what qualifies the bricks for what they do, for example "teaches at a high school", "manages sales in a shop") of bricks/persons combine in a more or less complex way. They thus give rise to inherited properties and behaviours. The assumption is that, for example, honest bricks/persons create an honest-emerging built society; or, with another example, unfairly advantaged bricks/persons create an unfairly built society. In more precise but technical terms, our Aristotelian Ur-philosophy, and the related sociological thought that is based on it, uncritically assumes an ontology formalized by "naive set theory".[6] This considers a set (in our case the society) as a variously complex and differently structured collection of simple objects, called elements or members of the set. And it analyses all the other non-atomic "social objects" (family, generation, social class, party, trade union, etc.) in terms of sets of natural or legal persons.

To this Aristotelian Ur-philosophy of things one then needs to add a Newtonian conception of space—for example, the house, the city, the region, the territory, the nation, the country, the borders, the land, the sea, the sky—and of time—for example, the days, the months and the years, the history, the tradition, the recurrences, the deadlines, the holidays—understood as two rigid and absolute reference frameworks or containers (not related to anything else), which are dynamic only insofar as they tend to an ideal definitive stability. Imagine a large box, space, in which the persons-bricks interact, in a linear and irreversible way, along the arrow of time. The fascist concept of "living space" ("spazio vitale") and the Nazi concept of

---

[6] See for example (Halmos 2017). Axiomatic set theory analyses sets on the basis of the relation of satisfaction of specific axioms.

"Lebensraum" are ideological aberrations of this Newtonian Ur-philosophy of physical space as geographic territory and physical time as a calendar.

As a whole, our Aristotelian–Newtonian Ur-philosophy of things, time and space puts all the emphasis on the concept of *action* as the essential point (the ontological variable, we would say philosophically) on which to press constructively in order to modify or improve the behaviours or properties (nature) of the elements/persons themselves and, above all, of their structural combinations, and therefore of the society they constitute. To simplify: according to this vision, society changes by operating on the *actions* of the natural or juridical persons that constitute it. The actions are therefore the point of pressure of the system on which to intervene in order to be able to manage, drive, or modify a society. From this, there follows a vision of law as a system through which one shapes the actions of agents-bricks (and their compounds) in time and space.

The metaphors of society as a body, organism, or system, or that speak of coordination, cohesion etc. are all based on this Ur-philosophy. One finds it in Menenius Agrippa's *Apologia* and later in Paul of Tarsus' advice to the Corinthians, as well as in the first pages of Hobbes's *Leviathan*. And from Weber onwards, the emphasis of sociological theorising on the concept of action indicates how the design of social architecture is still concentrated today on forming and directing behaviours by focusing only on actions and their effects as the entry points for any policy.

The crowning of the Aristotelian–Newtonian paradigm in sociological thought is the idea of constructing the social mechanism: atomic entities in their own right, thanks to their properties and behaviours, are combined into a structure that has its properties and behaviours, like an analogue clock. The construction of the desired mechanism, for properties or behaviours (in our case, a society), starts from the identification of the necessary and sufficient components needed to make it happen. If the mechanism does not work, or works in an unwanted manner, one may repair, modify, or add the responsible components, or the components that are necessary and sufficient for the solution, until they work as desired. The concept of "performance" and its quantitative analyses are the contemporary translation of this Aristotelian–Newtonian approach.

The Aristotelian–Newtonian paradigm has had its merits, but today it no longer responds to the needs of a mature information society, that is, a society whose members assume the digital as a foregone phenomenon (I will return to this concept later).

Since the twentieth century, the most formal and quantitative sciences—from mathematics to physics, to logic—confronted with more difficult conceptual challenges, have been forced to abandon the old Aristotelian–Newtonian Ur-philosophy or, if they still adopt it, they do so critically and with full awareness of its limits, essentially as a fall-back. Their "new" (but in fact now a century old) Ur-philosophy can be defined as *relational*. The problem is that our brain, our sensory apparatus, our languages and our Western cultures, by their nature, hypostatize (i.e. reify, or with a more intuitive term "thing-fy") the world, organizing it as lego: first there are things (nouns), then there are the properties of things (adjectives), and then behaviours of things (verbs). For example: Alice (what) writes (behaviour) with the blue pen (thing + property) on the white paper (what + property), and so on, for the rest of our experiential world. This is the way we are used to thinking. Our Aristotelian–Newtonian Ur-philosophy is so powerful because it is the codification of our deepest intuitions as intelligent mammals.

A *relational* Ur-philosophy uses sophisticated mathematical tools to overcome the obstacles of Aristotelian–Newtonian intuition and common sense in the various scientific fields. For example, relativity theory requires vector spaces, in which tensors are used to describe space and time in terms of four-dimensional spacetime. And category theory replaces set theory to uncouple the foundations of mathematics from the assumption of first elements understood as things, according to the Cartesian metaphor of the apples (elements) in the basket (together). The two examples are important, but they are also a bit disheartening. Because they are complicated and difficult. And if the request advanced in this chapter is to change the way we think politically, in the same way that we were forced to change our thinking on physics and mathematics, the suspicion that we are heading towards a resounding failure is justified. We can hardly understand how our democratic systems work. How can we therefore abandon such an intuitive and familiar Ur-philosophy in sociological reflection like the Aristotelian–Newtonian one, when we will have to dialogue with everyone (given no-one can exist outside of politics, as discussed), but without necessarily relying on a common conceptual vocabulary, especially when human reflection focuses on its conceptual and factual artefacts, such as society, economics, jurisprudence, and politics, which, by their nature, invite us to linger in a "natural" way of thinking? Society not only interprets itself in terms of "lego", it also builds itself in terms of "lego". Changing both trends seems a titanic effort destined to fail.

It must be admitted that the abandonment of an Aristotelian-Newtonian Ur-philosophy is a really difficult conceptual transformation, much more difficult than accepting that the earth is not flat, or at the centre of the universe, or that each of us is in large part a field of forces. The phenomena investigated—in our case, society and politics—impose a paradigm shift in a much less steady way (in terms of less intractable problems to be solved), and with much weaker standards (in terms of evaluation of the solutions). In other words, the Aristotelian–Newtonian Ur-philosophy is natural, intuitive, familiar, does not easily show its limits, and has worked in the past. The alternative is untested, counter-intuitive, unfamiliar, it is not how we conceptualise the world and our societies in it, or how we go about designing and constructing them, and does not really seem to be forced upon us by the nature of the problems with which we are dealing. It is going to be a hard selling.

An excellent example of this inability to think outside an Aristotelian–Newtonian paradigm is provided by Margaret Thatcher, not by chance an Oxford graduate in chemistry with a specialization in crystallography (few other scientific areas appear more Aristotelian–Newtonian). The "lego" model—which, I repeat, is now inadequate but difficult to replace—is evident in her famous interview from 1987:

> [...] there is no such thing as society. There are individual men and women, and there are families. And no government can do anything except through people, and people must look to themselves first. It's our duty to look after ourselves and then, also to look after our

neighbour. People have got the entitlements too much in mind, without the obligations, because there is no such thing as an entitlement unless someone has first met an obligation.[7]

Note the admission (for political rhetoric) that the family should be considered to be a basic element of society. In fact, it is contradictory: where does a family end? Do we include only parents and progeny, or even grandparents and aunts and uncles? And the cousins? And on what basis do we admit "the family" and not, for example, a group of families who are related to each other and represented by a village? And why not admit the whole human family? A dangerous slippery slope for any coherent thinker who is not also a politician.

The same (unsatisfactory) Aristotelian–Newtonian Ur-philosophy—this time stated almost literally, given the etymology of economics as regulation (*nome*) of the house (*oikos*)—is evident in Thatcher's simplistic conception of politics and the economy:

Any woman who understands the problems of running a home will be nearer to understanding the problems of running a country.[8]

Aristotle would have been happy with this statement. But we are not. Because today it is virtually impossible to understand accurately—and even more so, to manage successfully—in a simple framework of home government and ordinary Aristotelian–Newtonian insights, phenomena such as the purchase of their own shares ("buybacks") by a company; a policy of negative interest rates (a tax on owning money, to use Gesell's expression), supported by an inflationary monetary policy; the popularity, in recent times, of negative-performing government bonds, such as those issued by Germany; the fact that austerity is better exercised when it is possible but unnecessary, that is, in moments of economic growth, and not when it seems necessary, that is, in moments of crisis, when it is damaging; the goodness of a minimum degree of inflation; or populist and self-destructive phenomena of democratic implosion, like Brexit, the Trump presidency, and the success of populist parties in the Italian elections of 2018.

Society is not lego, and politics or a nation's economy cannot be understood in terms of mere management of household affairs. Thatcher was wrong. It is as if the CERN wanted to use only Newtonian physics to understand the behaviour of subatomic particles. The point is not that Newtonian physics does not work, but that it no longer works in this case, and that this case is now the more fundamental one.

In order to cope with the new challenges posed by mature information societies, where well-being is higher and more widespread than in the past (and compared with other developing societies), and the degree of complexity and interconnections is now profound, political thought must take a step forward and update the

---

[7] Interview 23 September 1987, cited in Douglas Keay, *Woman's Own*, 31 October 1987, pp. 8–10. https://en.wikiquote.org/wiki/Margaret_Thatcher

[8] BBC (1979), cited John Blundell, *Margaret Thatcher: A Portrait of the Iron Lady* (2008), p. 193. https://en.wikiquote.org/wiki/Margaret_Thatcher

common-sense intuitions espoused by the Aristotelian–Newtonian paradigm. But what concept can today replace the main one of a social *thing*?

Almost a century ago, Cassirer identified the end of what I have defined here as the Aristotelian–Newtonian paradigm in the transition from the centrality of the concept of substance (things) to the centrality of the concept of function (relations) in mathematics and physics (Cassirer 1923). He was right, and the next step is simple: a function is only a special kind of univocal relation between input and output.[9] It is therefore a matter of appreciating the possibility that it is not the concept of "thing", but that of "relation"—which refers to what constitutes all things and connects them among themselves—that can play a foundational role in the political thought of the twenty-first century.

We saw the difficulties, but there are also good reasons to be optimistic about the conceptual feasibility of this paradigm update. The conceptual vocabulary of relations is sufficiently rich, semantically, to allow us to express everything we want to express in the political vocabulary of things, their properties, and their actions. In more precise terms, the concept of relations is powerful enough to define all the necessary ontology.[10] This semantic equipotency makes possible something far more important than a mere translation exercise. It has the enormous advantage of moving and expanding (the dual movement is crucial) our focus first on the *analysis and design of relations*, rather than on the realization of specific actions or interactions, as the main point of pressure on which to operate to try to improve a society in a lasting and not ephemeral way. In simple terms: economics, jurisprudence, sociology, and above all, in our case, politics, become relational sciences of the links that make up and connect the *relata* (not just people, but all things, natural and constructed, and therefore their environments and ecosystems), even before being behavioural sciences studying the nature and actions of those special entities (that are natural and legal persons understood as things). In this, Hegel and Marx were perhaps prescient when they put the accent not on people themselves, but on the dialectical relationships *between* people.

This shift in conceptual paradigm changes the implicit operating model, which is no longer that of the Aristotelian–Newtonian mechanism, rather rigid and restrictive, but that of the force field or relational network, much more flexible, inclusive, and unbounded. In a network, nodes (including all people, but not only) do not pre-exist to be connected by relations, as is the case for the lego bricks or the components of a mechanism. Rather, they are the relations that make up the nodes, in the same sense in which the roads constitute the roundabouts. Therefore, if the properties

---

[9] Here, "relation" is to be understood in the logico-mathematical sense, as anything that qualifies every thing—human, natural, artificial—individually (e.g. Alice is unmarried, which is a un*ary* relation) or not individually (e.g. Alice and Bob are married, which is a bin*ary* relation; or Carol is sitting between Alice and Bob, which is a tern*ary* relation; and so forth for any n-*ary* relation).

[10] All entities are reducible to the totality of their properties, and all properties are reducible to n-ary relations, so all entities are reducible to the totality of their relations. Behaviours and changes in properties of entities are then reducible to state transitions, and the latter are reducible to transitions from one set of relations to another. In short, one can use the vocabulary of relations to speak of entities, properties, actions, and behaviours—and that is all that is needed.

or behaviours of the nodes-entities can be improved, it is on the nature and the number of the relations that constitute them that we must intervene. The new model, placing the relations at the centre of the socio-political debate, is more easily able to include in its analysis *all* the entities (relata), not only persons, but also the world of institutions, artefacts, and nature.

We know that things are discrete and can easily be grouped in separate sets. For example, we can group the set of all Italian citizens, the set of all French citizens, and the set of all citizens with both nationalities. Venn diagrams are popular for this reason. But social relations tend to be intertwined and continuous, with varying degrees of intensity, from weak to strong. In our example, we may be better off by speaking of Italian citizens who have relations with French citizens and vice versa in a variety of ways, i.e. relations that are more or less intensive, superficial, fruitful, frequent etc. As a consequence, in a "relation-oriented" and not "thing-oriented" policy, it is no longer the quantifiable amount of "performance" of things that is the main parameter of evaluation, but the degree of solidity and resilience of the relations that constitute things and bind them together, citizens included. When today we observe that, in some European countries, for example, the financial and political crisis has been addressed thanks to the efforts of families or social institutions, what we are actually saying, looking more carefully, is that it is the social network that today is making possible and less traumatic the transition from an industrial country (production of things and quality of things) to a country with a green and blue digital economy (production of services-functions and quality of experiences). This is not at all to contradict the phenomenon of globalization. On the contrary, a relational and not "substantial" (thing-oriented) view of society explains the current tendency of politics to become global and cosmopolitan, more based on diplomacy (a coming together of relations) than on war (a clash of things) according to a *reticular philosophy*.

This paradigm shift, which has been necessary since the rise of information societies, implies the abandonment not only of an Aristotelian ontology of the primacy of things, but also of a Newtonian ontology of space and time as rigid containers, within which things are positioned, move, interact, and change. Let's see how.

A network is a logical space, not a physical one, in which distances are measured with metrics that are not Euclidean. With an elementary example: in chess, the distance between a pawn and the queen is symmetrical in the Euclidean sense, for example 10 centimetres from the pawn to the queen and therefore from the queen to the pawn. However, it is asymmetric in the logical sense, for example a step from the queen to the pawn, but three steps from the pawn to the queen. Still in chess, the diagonal is necessarily longer than the column from a Euclidean point of view, but on the chessboard it has the same length in terms of number of squares, and therefore the king takes the same number of steps in covering both. In our case, with the arrival of the Internet, the space of politics (a relational and therefore logical space) no longer overlaps, indistinguishably, with the space of geography (a "substantial" and therefore physical space) of national sovereignty. This has been the case for a long time, since the old Westphalian identification of legal space with political space. On the contrary, the space of politics becomes the spatiality of social relations, including those of strength. The old concept of a "zone of influence" already antici-pates this idea in part. For example, the Mediterranean nature of Italy is above all cultural (i.e. relational), not merely geographical; likewise, Denmark is a Scandinavian country; and Spain can be as Mediterranean as Greece. This is why the EU should allow the expulsion of European member countries that do not respect agreements and shared values, and drop the geographical clause that pre-vents a non-European State from joining the European Union.[11] More Europe also means having the courage to abandon the twentieth-century geographical space, on which the EU was founded, to adopt a relational spatiality, making possible the exclusion of European countries that repeatedly deny the values of the EU, because geography is no longer sufficient, and the inclusion among its members also of countries not belonging to the continent, but which respect and promote its values, because geography is no longer necessary. From this new perspective it would be very reasonable to think of Canada, for example as a possible member of the EU, as has already been done in the past.[12] If this relational approach seems counterintui-tive, consider that it was already adopted with Cyprus, a State that, in terms of Newtonian space, geographically belongs to Asia, but which rightly entered the EU in 2004 on the basis of a spatiality made of historical–cultural relationships.

Similarly, political time takes care of the temporality of relations. For example, something becomes possible only after something else has happened: a concrete discussion of the feasibility of Eurobonds is conceivable only after the approval of the German government, in terms of the logic of chronological relations (before, during, after), and not of calendar year or calendar (absolute dates and times). And intergenerational relations are no longer relations between lego-like Aristotelian–Newtonian persons, but relational ties between node-like persons, something the vocabulary of politics describes as "social fabric", a crucial concept on which we need to pause for a moment.

To introduce this idea, it is useful to start from another version of the quotation from Thatcher we have already seen, on the sole existence of individuals and the non-existence of society:

> A transcript of the interview at the Margaret Thatcher Foundation website differs in several particulars, but not in substance. The magazine transposed the statement in bold, often quoted out of context, from a later portion of Thatcher's remarks: "There is no such thing as society. There is living tapestry of men and women and people and the beauty of that tapestry and the quality of our lives will depend upon how much each of us is prepared to take responsibility for ourselves and each of us prepared to turn round and help by our own efforts those who are unfortunate".[13]

---

[11] Article 49 (formerly Article O) of the Treaty on European Union, or Maastricht Treaty, states that any *European* country that respects the principles of the EU may apply to join. A country classifies as European "subject to political assessment" by the European Commission and more impor-tantly—the European Council. This geographic membership criterion was later enshrined in the so-called Copenhagen criteria.

[12] See https://mowatcentre.ca/canada-should-join-the-eu-sort-of/

[13] See https://en.wikiquote.org/wiki/Margaret_Thatcher

In a "tapestry" the fabric is woven in blocks of coloured weft threads, which are beaten down very tightly on the warp threads, producing a picture or pattern. When the work is finished, the warp threads are hidden. Weft and warp are sets of threads. Each thread is individual and the figures in the tapestry (and the tapestry itself) emerge from their intertwining. So, Thatcher was right in the choice of her conclusion: her likening of the "social fabric" to a tapestry is correct, if one looks at the internal coherence of her ideas. But a fabric does not necessarily have to be "woven" like a tapestry, it can also be knitted (a word that comes from "knot", which clearly relates to network), like a blanket. And in this case, it is a fabric formed by a number of consecutive rows of intermeshing loops. The loops do not pre-exist the fabric, but co-exist with it because of the common thread. Thus, Thatcher was wrong in choosing the premise: because the social fabric is a lot like a knitted blanket, not so much like a woven tapestry.

Finally, the *personal* fabric is the "inter-temporality" of an individual life, that is to say, the fact that human existence, individual and social, is like a knitted thread, whose loops must relate correctly with each other according to a coherent design. For example, if one invests in higher education one should then find a place in society to work. There must be inter-temporal links that give meaning to paths, trajectories, expectations, individual and social human projects (more on this later), and so on. Politics must know how to take charge of the "inter-temporality" of people's lives and of the intrinsic relationships and connections between the phases of human existence, addressing not only individuals' interests but also their hopes, as we shall see shortly.

The social fabric is the systemic interdependence between the all the individual personal fabrics, of which I have just spoken. Today, in many countries, this social fabric is also interpreted in terms of an information society. And in some cases, we can already start talking about mature information societies. It is a concept that I have already used and that we shall need in the rest of the chapter. The time has come to analyse it in detail in the next section.

12.3   The Idea of a Mature Information Society

We are so familiar with talk of "*the* information society" that we sometimes forget there is no such thing, but rather a multitude of societies, different from each other, some of which may qualify as information ones, in different ways and degrees. We should really speak of "information societies" without a "the" but with an "s", and ensure that our generalizations are not so generic as to apply to any of them, indiscriminately, obliterating every salient distinction. Just to be clear, there is always a level of abstraction at which something is like anything else: the moon is like your umbrella, which is like a pizza, because they are all singular objects that exist and look round, for example. The point is not being smug about one's own acrobatic equations (x is like y which is like z) but being critical in checking whether the level of abstraction at which the equation is drawn is the fruitful one to address the question one is trying to answer. All this should clarify why, once we have acknowledged that there are many information societies that are all different from one another, it still makes sense to compare them in terms of relevant criteria and why, more specifically, it is important to understand what it means for an information society to be more or less mature than another.

Maturity is a matter of people's expectations, not just technological or economic development. Let me first explain the difference with a concrete example, and then with an analogy.

The Organisation for Economic Cooperation and Development (OECD) collects many statistics that are useful for evaluating the stage of development of an information society. Because of the OECD's remit, they mostly refer to technological advancement. These statistics are clustered around four main areas of development: broadband and telecoms, the Internet economy, consumer policy, and digital government. Together, these areas may seem to provide a good snapshot of what it means for an information society to be mature. Indeed, this is a common approach. And yet it is far from satisfactory. Take the percentage of fibre connections in total broadband among countries reporting fibre subscribers (see Fig. 12.2). Information updated to December 2014 shows Japan, Korea, and Sweden in the first three places, something that is not surprising; but it also ranks Turkey way above the Netherlands and the United States, while at the bottom we find countries such as Finland, Germany, and Ireland, which one may consider rather mature information societies.

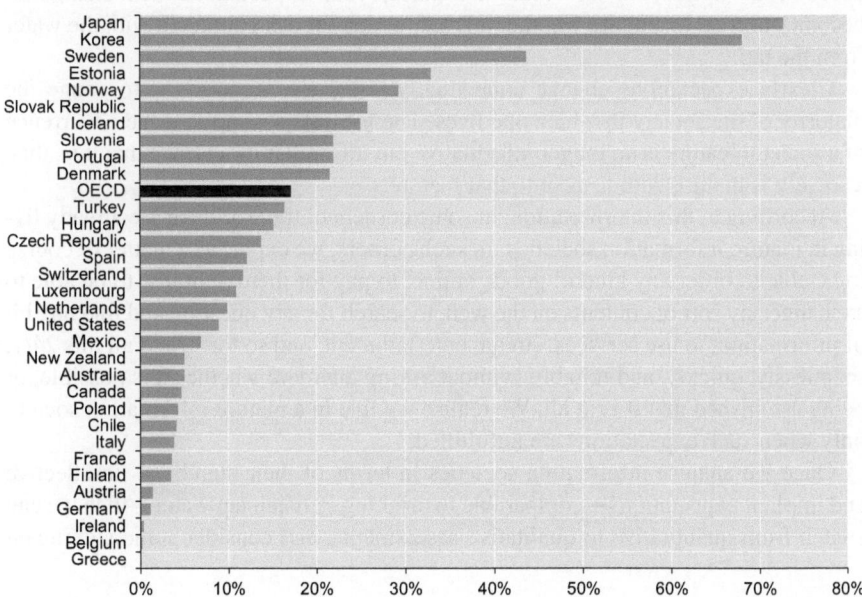

Fig. 12.2 Percentage of fibre connections in total broadband subscriptions, December 2014. (Source: http://www.oecd.org/internet/broadband/oecdbroadbandportal.htm)

The percentage of fibre connections clearly does not provide a good criterion. Similar criticisms could be levelled at all the other statistical data collected by the OECD. The trouble with technological and economic approaches—think of connectivity, Internet penetration, number of computers per household, government open data projects, e-health services, usage of social media, ICT investment per capita, and so forth—is that they capture only some of the conditions that facilitate the development of an information society. Such conditions are important, because they represent very helpful and significant affordances. Yet they are neither individually necessary nor altogether sufficient in themselves to explain in what sense and why an information society may be more mature than another. Something crucial is missing, namely people's *expectations*. Consider now the following analogy.

When you are in a hotel in Paris, you rightly expect the water in the bathroom to be drinkable because France is a "water mature society". In fact, you do not even think about it. There is no need for the hotel to advertise the safety of its water, nor for you to ask at the reception whether the water is drinkable. France is a "water mature society" not just because of its water system, but because people living there treat drinkable water as something ordinary, non-informative, a matter of fact that lies in the background. It is part of life, of what you implicitly and unreflectively expect water to be like in Paris. At the same time, we all know that drinkable water is not a trivial matter. According to the World Health Organization and UNICEF Joint Monitoring Programme, in 2015 nearly 700 million people—a tenth of the world's population—did not have access to safe water.[14] So if you take a more adventurous holiday in an unfamiliar place, your expectations will change. It becomes normal to enquire whether it is safe even to brush your teeth with the water from the tap.

Clearly expectations change contextually. They are a good way to gauge the maturity of the society in which one lives. The formula is simple: if the occurrence of a societal feature is no longer informative but it is rather its absence that it is, then a society is mature with respect to that feature.

According to this interpretation, in some corners of the world we are already living in mature information societies. In such corners, we expect as a matter of course to be able to order any kind of goods online, to pay for them digitally, to be able to exchange any sort of contents on the web, to search for any question and find any bit of information, to use services, stream entertainment, and so forth, and all this 24/7, seamlessly, quickly, and reliably, without asking anymore whether it is possible, or being astonished that it is at all. We realise we live in a mature information society only when such expectations are unfulfilled.

Once we analyse information societies in terms of their members' unreflective and implicit expectations—comparable to drinking Parisian tap water—then we can switch from quantitative to qualitative assessments, and consider some significant

---

[14] Source: Progress on Drinking Water and Sanitation, 2015 Update and MDG Assessment, http://www.who.int/water_sanitation_health/monitoring/jmp-2015-update/en/

consequences. In this context, three seem to be more important than others; in no particular order of importance.[15]

First, *education*. One can dismiss the myth that "these days young people all know how to use [add your preferred technology here]". Young people are different from old people because of what they take for granted, i.e., their implicit expectations, not because of their innate skills, which, since Lamarck was wrong and Darwin was right, are no better than the skills of any other member of the society in question. Children have no innate knowledge of how to use touch screens. Yet, they are astonished if screens do not somehow respond to their touch, because that is their default expectation; what they grew up with. Compare this to an automobile mature society. Precisely because it is not a matter of knowledge or skills but implicit assumptions and quiet expectations, Alice or Bob may live in an automobile mature society in which ordinary life is full of cars, parking spaces, traffic lights, petrol stations, motorways and so on, and all this working in some ordinary way, even if they do not know how to drive a car, have no idea how a car engine works, or how a car can be fixed. They may take the availability of public transport for granted, for example, and have no driving license. So not only does the "young generation" have to learn—as does any other generation—how to use specific technologies. It is also *irrelevant* whether it does, in order to understand whether such a generation lives in a mature information society. We rather need to check people's expectations. This is one of the many reasons why education is important: it makes one aware of one's own implicit assumptions and expectations, and of their justification, reasonableness, and historical determination.

Second, *understanding*. Expectations may be blinding towards alternatives. As with the water analogy, it is crucial to know what one may rightly expect from one's own society—drinkable water or good Internet connectivity—while knowing that this is probably very far from, and way more privileged than, what millions of people enjoy in many other places. If you expect your bank to provide a scanning app to deposit cheques automatically online without walking into a branch and filling out a form, then you can rightly complain if it does not, while still being able to appreciate the fact that many people do not even have access to an ATM. More generally, a society implements a particular right increasingly well the more its lack becomes informative.

And finally, *innovation*. We saw that expectations determine what is ordinary and what is extraordinary, what is normal and what is abnormal. In short, they determine what is informative[16]: being told that the water in your hotel in Paris is *not* drinkable (e.g. because of some work in progress) would be very informative, as would be having access to a high-speed Internet connection in a remote, beautiful island in the middle of nowhere. So, innovation in a mature information society is culturally (mind, not technologically) more difficult to achieve than in an immature one. Because once people start taking for granted cars or TVs, for example, it does not really matter how much these technologies change, maturity has been reached, in

---

[15] For an overview see (Floridi 2009).

[16] And hence what needs to be designed, see (Floridi 2008, 2011, 2017).

terms of more of the same. Likewise, once people are used to living "onlife"—i.e. in that liminal estuary represented by the mixing of online and offline experience (Floridi 2014a, b)—more digital products, goods, services, or in general more affordances will not make that society even more mature.

Information societies are maturing all over the world. More will appear in the future. In terms of expectations, similarities therefore will increase. To paraphrase Tolstoy, all mature information societies are alike in terms of people's expectations; each immature society is immature in its own way. So, the next stage in the development of information societies, be this in 10 or a 100 years, will not be a further maturation of their inhabitants' expectations about their digital affordances, it will be an unprecedented and unforeseen transformation altogether, for which the digital will have become an implicitly expected backdrop. This might be hard to imagine, at a stage in which we are still going through our information revolution (Floridi 2014a), but then history has a tendency to upgrade our imagination on a regular basis.

12.4   The Idea of a Human Project

By "human project" I mean the kind of life and society we would like to achieve. In a more simplistic way, it is what political parties, often without critical conscience, try to summarize in their electoral slogans, for example "For the many not the few" (British Labour Party 2017), or "Building a country that works for everyone" (British Conservative Party 2017). In a more analytical way, the human project is the form of human life—programmatic in its various individual, collective, private, and public manifestations—that a society presents and promotes from time to time as desirable, at least in theory or implicitly, and depending on historical moments. Perhaps the most suitable philosophical term to describe the concept of human project is the German term of humanity's social *Lebensform*.

It is plausible that each human project, at every stage in human history, is not entirely feasible, or is only minimally feasible, and therefore should be understood only as a regulatory ideal,[17] towards which to aim. Despite this limitation, two crucial observations remain correct.

First, each society incorporates its own human project, no matter whether this is only implicitly or explicitly pursued, whether it is coherent or contradictory (for example, when it comes to promoting several projects that cannot be reconciled with one another), pragmatic, realistic, or utopian. This happens for two reasons. Because individuals get together, voluntarily or not, on the basis of a shared purpose—the human project—be this positive (as in Plato), in order to achieve a higher degree of trust, coordination, and collaboration; or negative (as in Hobbes), in order to achieve a lower degree of distrust, conflict, and insecurity. The second reason is because the very absence of a human project is *itself* a project. We are back to the

---

[17] The implicit reference to Kant is meant.

relational nature of phenomena that also absorb their negations. Not having a project does not mean you are doing without one, but rather that you have opted for a bad project, underdeveloped and uncontrolled. It follows that a society without a human project does not exist. There are only societies with human projects that are more or less good, achievable, or compatible with each other.

Second, although every society usually tends to absolutise its human project as *unique* (there is only one, its own), *eternal* (its own is always valid) and *universal* (its own is valid everywhere), in reality there is no single human project, but as many human projects as there are societies, states of societal evolution, and historical circumstances in which they are found. This *pluralism* is not *relativism*, as if one were saying that every human project is necessarily as good or bad as any other. In reality, it is a matter of adopting a serious and relational way of describing the plurality of the projects in question, as made possible also by what has already been achieved, and therefore known, and by what has not been realized, but it is conceivably achievable. The human project described by Cicero in *De Republica* is very different from the one described by Tocqueville in *De La Démocratie en Amérique*, and neither is easily applicable today to the information society.

Among the various factors that explain the sense of radical transformation and uncertainty characterizing our time there is, above all, the implicit perception of the absence of a human project in the information societies that are maturing before our eyes. The metaphor is that of ever-faster traveling in a still unknown and sometimes obscure direction. The problem is exacerbated by the fact that we do not have a human project for the digital age (to be precise: we obviously have a project – as absence of project is itself a project – it is just not planned). However, we do have a postmodern starting point, in the sense of an incomplete meta-project shared by the industrial and post-industrial consumer society, which today characterizes many advanced economies. The old project is dying but the new project struggles to come to life, to paraphrase Gramsci. Both terms, "meta-project" and "incomplete", need to be clarified.

The postmodern metaproject consists in the fact that the information society, like the consumer society, pursues the human project to make the various individual human projects possible and compatible with each other. In other words, in the best of cases, today the human project is reduced to the support of individual projects (aspirations, hopes, plans etc.), that is, to the social project to make the various individual projects feasible and compatible with each other. To exemplify, we do not pursue a "happy society", but rather a society in which every individual has the opportunity to pursue his or her own happiness, provided this is not at the expense of others. The examples can be multiplied: we do not pursue a rich society, but a society in which every individual has the opportunity to get rich within the limits of legality; not a healthy society, but a society in which every individual has the opportunity to live and take care of him or herself in a healthy way within the available constraints; and so on.

The meta-project is clearly *liberal*. The purpose of the State is centred in defending and promoting the rights of each member of society, in a mutually compatible way. And the mechanism on which this relies is that of the "spontaneous" emergence of the desired social-relational properties, starting from the realization of the individual relations that are supported. It is an approach still based on the "lego" model we already encountered. In the previous example, allegedly, a happy society would spontaneously emerge from the happiness of its members. Economically, this emergentism goes well with *liberalism*: the State ensures a free market in which individuals can own, produce, and trade economically, within the limits of legal compatibility. In some cases, ethical liberalism and economic liberalism end up supporting political *libertarianism*, which promotes the maximum reduction of the functions of the State in favour of the freedom and responsibility of individuals.

The liberal and postmodern metaproject is incomplete because it focuses only on the interests and hopes of the individual, or at most of the person, including the legal person (think of corporate taxation), but does not provide, nor does it mean to provide, programmatically, an indicative framework on the kind of society that one would like to build together, and for which coordination of the efforts of many, if not all, is needed. I will return to this second point in the next section. Here it is worth stressing that, in the past, starting from the twentieth century, the incompleteness of the postmodern meta-project was made less evident by the great historical disasters of the two World Wars and by the ensuing reconstructions, by the Cold War, and by political and religious ideologies. Whenever we had to fight together against something for something—for better or for worse—or to build or rebuild together what we inherited from this fight, or whenever we adopted a collective ideological or religious faith, in all these and similar cases the postmodern metaproject was supported *externally*, by other social or community projects, which hid its incompleteness. The great movements for various human rights, the pacifist and ecological movements, for example, have provided the social component to the postmodern human project, which otherwise would long have remained limping on the single leg of the individual human metaproject. In the best of cases, these external social projects have been "included" in the human project, providing it with the non-individualistic component. Think of the work of Martin Luther King in the United States, the fall of the Berlin Wall and the reunification of Germany, or the end of Apartheid in South Africa. The same happened for political and religious ideologies. Fundamentalism and populism are also answers addressing, implicitly and uncritically, the incompleteness of the current human project.

Today, the gap between social projects and political projects is extreme, and the former can no longer hide the incompleteness of the latter (using a British example, Cameron's "Big Society", with volunteering replacing government services while hiding a paucity of political ideas, did not last long and was a failure). The social project, whatever it is, is no longer part of the political project. Quite the opposite, it often distances itself from the political project in an anti-political way, falling into the negative dialectic described in the introduction. The proof is that the world of volunteering and therefore social commitment grows together with disenchantment for political commitment, and its refusal. For example, according to the latest ISTAT data, in 2013 in Italy 6.63 million people (12.6% of the population) volunteered their time and work for free and for the common good. In the light of what I have argued, this is not a contradiction, but a consequence of an incomplete human project: politics has not taken on a social human project, and this need, which goes beyond individual human projects, is otherwise met, outside of politics.

This generates three risks. We have already discussed the first one. Community activism, detaching itself from the human project, risks leaving it unlimbed and limping. The second is the double illusion that community activism can somehow compensate for the absence of a *social* human project—as opposed to an individual human project—and that politics can not only be left limping but continue so without any negative consequences. The third is that community activism is confused with the social human project and tries to replace it, through movements that claim to be political, but do not intend to do politics positively, because they fail to recognise that *good politics is the properly regulated evolution of community cooperation.*

All this leads to a crucial question, which is essential if we are even to hope to be able to outline a good human project for a mature information society: if it is possible to adopt not only an individual metaproject, but also a social project—and this conditional is not rhetorical at all—is it possible to do so today without falling into a right-wing or left-wing ideology, or a religious one? In other words, is a *complete* human project possible, both as a metaproject for the individual and as a social project, that is, neither ideological nor transcendent? I believe the answer can be positive, but the room for manoeuvre is narrow. Let's examine it.

It is indicative that one never speaks of a centrist ideology. The centre of politics does not have its own ideology because, in the best of cases, it transcends the latter, adopting ethics as the main and superior guide. And in ethics—from Aristotle to Rawls—the end is always that of equilibrium and of a collaborative reconciliation of interests, rather than the imbalance of the confrontation of parts, in a zero-sum game. The centre does not promote or lead "political struggle", but creates political convergence; connects, does not disconnect; it does not quarrel, it argues. For this reason, the human project that we can hope to draw today can proceed socially and not only individualistically (and hence only metaprojectually), if it is pursued in an ethical-centric way, and not in an ideological way of left or right; and in an immanent and not transcendent way, staying within history and improving it from within, not coming out of it in a saving way, and rejecting it. That is to say that good politics will no longer take left vs. right seriously, but will concentrate on centrist alternatives that have more or less successful strategies to approach the human project. To be coherent, the ethics to be adopted will have to be inclusive of all those parts of the world and society inevitably ignored by the meta-project, that is, those parts that do not play an active role in presenting and managing their own interests and rights in the first person. It is one of the great lessons that political commitment can learn from the community commitment: the human project for the digital age and for a mature information society must include the "silent world": the marginalized, the disadvantaged, the weak, the oppressed, the past generations to be respected, and the future ones to be facilitated, the environment (natural and artificial), and that semantic capital formed by culture and memory. In other words, it must be an ethics of the interests of all the "patient" nodes (those who receive the effects of political action), and of the various networks that they form, and not only of the individual "agent" nodes, whose interests are already taken care of by the metaproject component, which knows their requests because they are presented explicitly and constantly. It will have to listen to those who are not heard by the metaproject.

As for the relationship with religion, the human project must support a secular and immanent society, while being fully respectful of the faiths that can not only cohabit but also flourish within it. The reasons in favour of a lay human project are many. Only a secular society can be coherent with the meta-project, which, to repeat, is a project to facilitate individual projects to the extent that they are mutually compatible. Only a secular society can be truly tolerant, that is, sincerely respectful and supportive of the great variety of individual human projects. And only a secular society can lack any interest in proselytism, and not fall into the temptation of *imposing* a specific vision (religious or otherwise) of the human project at the expense of other visions, or a specific evaluation of the world as comprising "we" and "they" (religious divide). The human project will need to be secular and lay because ethics can unite and support faith, but faith often ends up dividing and defeating ethics.

To sum up, the human project for a mature information society must first be ethical and then be political, and it will have to be made up of two components, one now classical, represented by the liberal metaproject that favours individual projects, and the other still to be built, which can also make social sense of the way we live together, as a community. The fact that today there is no serious utopian thinking shows that we have not yet developed the second part. To fill this gap, we need an important thing: a good ethical infrastructure that allows coordination and care of the social fabric. This is the topic discussed in the next section.

12.5   The Idea of an Infraethics

It is a sign of our times that, when politicians speak of infrastructure nowadays, they often have in mind information and communication technologies (ICTs). They are not wrong. And it is an old story. From success in business to cyber-conflicts, what makes contemporary societies work depends increasingly on bits rather than atoms.[18] Depending on their digital infrastructures, societies may grow and prosper. And it is ICTs that can also present a catastrophic weakness, in terms of cyber security and the vulnerability of our increasingly networked critical infrastructure. We know all this. What is less obvious, and philosophically more interesting, is that ICTs also seem to have unveiled a new sort of equation.

Consider the unprecedented emphasis that ICTs place on crucial phenomena such as accountability, intellectual property rights, neutrality, openness, privacy, transparency, and trust. These are probably better understood in terms of a platform or infrastructure of social norms, expectations and rules, that is there to facilitate or hinder the moral or immoral behaviour of the agents involved. By placing at the

---

[18] The implicit reference to Negroponte is meant.

core of our life our informational interactions so significantly, ICTs have uncovered something that, of course, has always been there, but less visibly so in the past: the fact that moral behaviour is also a matter of "ethical infrastructure", or what I will simply call *infraethics*.

The idea of an infraethics is simple, but the following "new equation" may help to clarify it further. In the same way as business and administration systems, in economically mature societies, increasingly require physical infrastructures (transport, communication, services etc.) to succeed, likewise human interactions, in informationally mature societies, increasingly require an infraethics to flourish. The equation is a bit more than just an analogy between infrastructure and infraethics. When economists and political scientists speak of a "failed state", they may refer to the failure of a *state-as-a-structure* to fulfil its basic roles, such as exercising control over its borders, collecting taxes, enforcing laws, administering justice, providing schooling, and so forth. Or they may refer to the collapse of a *state-as-an-infrastructure* or environment, which makes possible and fosters the right sort of social interactions. This means that they may be referring to the collapse of a substratum of default, accepted ways of living together in terms of economic, political and social conditions, such as the rule of law, respect for civil rights, a sense of political community, civilised dialogue among differently-minded people, ways to reach peaceful resolutions of tensions, and so forth. All these expectations, attitudes, rules, norms, practices—in short, such an implicit "socio-political infrastructure", which one may take for granted—provides a vital ingredient for the success of any complex society. It plays a vital role in human interactions, comparable to the one that we are now accustomed to attributing to physical infrastructures in economics.

The idea of an infraethics can be misleading, because, despite the economic analogy, an infraethics should not be understood in terms of Marxist theory, as if it were a mere update of the old "base and superstructure" idea. The elements in question are entirely different: we are dealing with moral or immoral actions and not-yet-ethical facilitators of such moral or immoral actions. Nor should infraethics be understood, conceptually, in terms of a kind of second-order or metaethical discourse about ethics, because it is rather the not-yet-ethical framework that can facilitate or hinder evaluations, decisions, actions, or situations, which are then moral or immoral. At the same time, it would also be wrong to think that an infraethics is either ethically *neutral* or simply has an ethical *dual-use*, because its dual-use is always *oriented*. If it were just neutral, this would mean that an infraethics would not affect either ethical or unethical behaviour, a mere logical possibility that is utterly unrealistic. In philosophy of technology, it is now commonly agreed that design—in any context, society included—is never ethically neutral, but always embeds some values, whether implicitly or implicitly. Yet this does not mean that an infraethics is simply dual-use, as if it could both facilitate and hinder morally good as well as evil behaviour in equal degree, depending on other external factors. The textbook example is the knife that can save a life or murder someone. And the trivial comment is that its use and hence moral evaluation depends on the circumstances. This is true, but insufficiently perceptive. Because not all knives are born equal. The very short, blunt, round knife that an airline provides to spread butter has a dual-use that is hugely oriented to fulfil a purpose that the butcher's knife can also fulfil, but much less easily. A bayonet has a dual-use only theoretically, because it is designed to kill a human being, not to cut bread. Likewise, every infraethics may be dual-use only in principle: in fact, if it is a good infraethics, it means that is oriented towards facilitating the occurrence of what is morally good. At its best, an infraethics is the grease that lubricates the moral mechanism in the right way and successfully. So, it is easy to mistake the infraethical for the ethical because, whatever helps goodness to flourish or evil to take root, it partakes of their nature.

As I mentioned in the previous section, speaking of the need for a human project that is not only meta-conceptual but also social, every society—be this the City of Man or the City of God, to put it in Augustinian terms—pursues its human project (even if only unconsciously) by adopting (even if only implicit) an infraethics, which can be more or less morally successful, and more or less evil-unfriendly. It follows that even an ideal society of angels, that is, a society whose nodes are all impeccable good moral agents, needs infraethical rules for coordination and collaboration. In other words, not even a society of angels can succeed if it is exclusively a libertarian one. It too needs a social project to support its development. Thus, James Madison was *partly* (more on this specification below) mistaken when he famously wrote that

If men were angels, no government would be necessary (*The Federalist* No. 51, 1788).

He was partly mistaken because he had a merely negative anthropology in mind—the one so well-articulated by Thomas Hobbes in his *Leviathan* and *De Cive* ("homo homini lupus") and never revised nor criticised by John Locke—and an atomistic view of society as a mere aggregate of individuals (recall the Aristotelian–Newtonian Ur-philosophy). Yet even a society of angels would still need some form of government, and hence an infraethics, to coordinate its good deeds, set common goals, evaluate the degree of success in pursuing them, and rectify the course of actions as a group, if necessary. Because "good" can always be "better" and "we the people" is not equivalent to a mere aggregate of all the Alices and Bobs in the world. An arch is not only a pile of stones. There is a moral goodness that is entirely social and does not emerge merely from individual moral goodness. Because goodness is also a matter of ambitious agency: what "we the people" can do and hope to achieve together, as opposed to what Alice and Bob could ever do individually. Angels would still need an infraethics to organise a party, or to push-start a car. It is not always true that every little effort helps: an angel attempting to push the car on its own will only waste its time and effort, completely. A multi-agent system—many angels working together to push the car successfully—needs coordination and control if it is to achieve anything. We should also regard evil is a matter of opportunity costs (not just bad deeds), that is, what *could* have been done that wasn't. Without a system of governance, the angels will miss performing many good deeds that are only available to them as a group. This cost can be very high and morally negative in any society.

I specified above that I take Madison to be only "partly" mistaken about his posi-tive assessment of angels as requiring no governance because that sentence should be read within its context, which states that

> If men were angels, no government would be necessary. *If angels were to govern men, neither external nor internal controls on government would be necessary* (*The Federalist* No. 51, 1788, my italics).

The part in italics shows that Madison was actually referring to the need to structure the government with checks and balances (that is, with external and internal con-trols). So, one may read him more charitably, not as saying that any government, or any infraethics, would be unnecessary—as stated in the first sentence—but rather as saying that one designed on the basis of an angelic anthropology would be. That is, he might be interpreted as arguing not that rules for coordination would be unneces-sary, but that special constrains on the application of these rules would be unneces-sary if we were angels, because those governing and those governed would behave according to the proper application of the rules all the time. With an analogy, he might be read as saying that, if all men were angels we would still need driving rules to coordinate driving behaviours, but no police to enforce them.

Insofar as Madison was mistaken—the first sentence of the quotation above defi-nitely is, and it is often interpreted by itself as meaning what I took it to mean above, as if every law and social regulation were based only on the dialectic between "crime and punishment"—it would also be wrong to dismiss the crucial importance of an infraethics not only from a libertarian but also from an anarchist perspective. In this case, the reasoning shares the premises and draws a different conclusion: if men were angels they would need no government, but men (sometimes) are angels, and so (sometimes) they do not need government. The spontaneous emergence of the morally good is therefore (erroneously) assumed as both natural and uncontro-versial by libertarian and anarchist alike. Yet the truth is that without an infraethics to begin with (i.e. internal controls), and then the issuing of good governance that supports it (i.e. external controls), not enough moral goodness could ever be achieved individually. A multiagent system like a whole society needs its own organisation and governance, precisely because it is not an old Aristotelian–Newtonian cuckoo clock.

If we now return to the *oriented dual-u*se of an infraethics, one may argue that a society of Nazi fanatics could rely on high levels of trust, respect, reliability, loyalty, privacy, transparency, and even freedom of expression, openness, and fair competi-tion, without being for this any less evil. Clearly, what we want is not just a success-ful framework of facilitations and constraints provided by the right infraethics, but also a coordinated cohesion between them and morally good values, such as human and civil rights. This is why a balance between security and privacy, for example, is so difficult to achieve, unless we clarify first whether we are dealing with a tension within ethics (security and privacy as moral rights, i.e., both understood as "water" in the earlier analogy), within infraethics (both are understood as not-yet-ethical facilitators, i.e. as part of the pipework), or between infraethics (security intended as facilitator or "pipe") and ethics (privacy intended as a value, or "water"), as I suspect to be the case.

The right sort of infraethics is there to support the right sort of values (that is, axiology). Designing it, maintaining it and keeping it updated is one of the crucial challenges for our information society. It is also one of the reasons why, in terms of innovation, our age is the age of *design,* even more than an age of discoveries or inventions. Clearly, when politicians talk about "infrastructure" nowadays, they often have to deal not so much with bits and atoms, but rather with the infraethics and the values it supports. It is mainly working on these last two factors that politics can best support the right human project at the right time—for a mature information society.

12.6   Ideas for a Mature Information Society

We have seen that political thought should move from a "substantial" to a "relational" approach, from mechanisms to networks. This means thinking of politics as a *science of relations* and as a guide and management of the *ratio publica* (see below) even before the *res publica.* The new relational paradigm helps us to understand how an information society, which is mature in terms of its socio-cultural expectations, can articulate and pursue its own complete human project—that is, both an individual human project as a meta-project for individual projects, and a social human project, for group projects—using the right infraethics to organise itself and realise it. All this makes possible, and at the same time requires, good ideas for a better politics. This is both in the sense of positive conditions of possibility, which aim to draw and then build what is or should be a good democracy for a mature information society, but also in the sense of negative conditions of possibility, which reveal the presence of bad politics, which hinders the construction of what is or should be a good democracy.

In this section, I present some of these ideas, those that today seem to me to be the most important. They can be read as conceptual explanations or logical consequences of a single premise: what a good politics for a mature information society is—that is, a politics that intends to pursue a complete and ethically desirable human project, through an effective and sharable infraethics.

I have tried to facilitate the task of the reader by schematically separating the various ideas and numbering them, so that it may be easier to agree or disagree with each of them. I have italicized some key concepts when they are introduced for the first time before being discussed or explained. And I tried to make the text readable on two levels. The first level is a network that simply connects every numbered idea, readable as a node, while ignoring the paragraphs below, which represents a further analysis. For those in a hurry, it should be enough to read just the numbered phrases. For those who have time and patience, the second level is more in-depth and sequential, and requires a non-reticular reading.

1. A society is the totality of the *relations* that constitute it.[19]

   This is because a society is formed, and not merely composed, by many individuals, who are not like stones collected in a pile, but who interact, coordinate, and change.

2. A *good society* is a *tolerant* and *just*, and therefore *peaceful* and *free,* society.

   These four moral values, presented in order of logical precedence, are essential. They refer to the four conditions identified by Locke (tolerance is the foundation of peace), by Mill (tolerance is the foundation of freedom) and, between the two, by Kant (justice is the foundation of tolerance). However, in this chapter I argued that tolerance and justice have this logical order (tolerance has priority over justice), even if they are co-necessary.[20]

3. A *civil society* is organized into a political community, called a *polity*.

4. A *government* is the executive guide of the polity.

5. *Governance* is the activity of the government.

   Governance includes the design and management of social policies, with proper oversight, transparency, and accountability.

6. *Democracy* is the best way to create and maintain the governance of a polity.

   This is because democracy maximizes the just care and tolerant flourishing of individual, social, and environmental relations, paying attention to the satisfaction of the interests, needs, and reasonable hopes of not only all persons (both physical and legal) but of all related "things", that is, the human, natural, and artificial *relata*.

7. The best form of democracy is *representative*.

   This is because a necessary condition of democracy is the structural separation between *popular sovereignty* (those entitled to vote hold political power and can legitimately delegate it) and *political governance* (those who rule receive political power and can legitimately, transparently, and accountably exercise it, through revocable delegation). From this it follows that all forms of dictatorship—including that of the majority—spring from the self-legitimizing merging of *sovereignty* and *governance*, that is, between the possession and the exercise of political power. Every form of government and governance is fallible: sometimes they do not work, or they work badly. From this it follows that a representative democracy is preferable to a dictatorship not because it works better, but because it is much more *resilient*: when it does not work, it works much less badly than a dictatorship because it causes less damage and admits of change and repair.

8. Good democracy allows voters to choose between *real alternatives*.

   This means that the multiplication or superfetation of choices and the lack of real alternatives of content is a hallmark of any anti-democracy in any political regime. It reduces the space of political decision: voters choose between options

---

[19] The implicit reference (in disagreement) to proposition 1.1 of Wittgenstein's *Tractatus* ("The world is the totality of facts, not of things.") is meant.

[20] See also (Floridi 2015, 2016).

(as in a restaurant menu), but do not decide between alternatives (which restaurant to go to).

9. Good democracy offers the right *granularity of alternatives*.

   This means that the more we collect packages of choices (bundles) in individual blocks on which to ask to decide politically, the less good the democracy in question is. This is an argument in favour of a mixed electoral system, with some balance between majoritarian and proportional features, to reach the right level of granularity.

10. A good society requires a *good politics*.

11. Politics is bad when it does not allow change to an individual's *starting position*.

    The impossibility of modifying one's starting position constitutes another hallmark of the anti-democracy of a political regime, and it is equivalent to the reduction of space in the construction of the human project. Social mobility, for example, is a sign of good politics.

12. Good politics seeks to take care of the *prosperity* of the *whole society*, of *all* the people who belong to it, and of public and common goods, including natural and artificial environments, which belong to it or in which it lives. "All" here means, ideally, not only the society that expresses it, but the entire human society.

13. *Prosperity* is a relation that includes the protection and promotion of civil liberties, education, security, wellbeing, and equal opportunities. Following a relational and not "substantial" approach, arguing that good politics takes care of the prosperity of the whole society, of all the people who belong to it, and of public and common goods (including natural and artificial environments), means ensuring that politics is *reticular*.

14. The *ratio publica* is the totality of public, individual personal fabrics, the social fabric and the fabric of public and common goods.

15. Good politics is *textile* (fabric-like), as it takes care of the *ratio publica*.

16. Politics is bad when it tears the fabric of the *ratio publica*, failing to ensure a minimum level of decent life, individually and socially. For this reason, the violation of the dignity of the person or of groups of people constitutes another hallmark of the anti-democracy of a political regime, reducing the space within which one may flourish in a society.

17. Good politics is *universally participatory*.

    Good politics requires the input and active participation of all the components of a society, including industry associations, companies, and administrative structures. Good politics is successful only if there is the involvement of all the stakeholders, at all stages, from the initial brain storming and reflection, to the development of good ideas, to their discussion and implementation. Participation has no natural boundaries, but only pragmatic limits. This is why good politics is also inevitably *cosmopolitan*.

18. Good politics can be transformed into good governance only thanks to the positive support of the *public administration*. Failing to work in synergy with the public administration is not only a strategic mistake, because the public

administration knows the mechanisms and degrees of feasibility of political projects from within, it is also a mistake of perspective, because only the commitment of the public administration can guarantee the continuity and the final success of the projects even across several governments.

19. Implementing good politics together with the social partners and the public administration means drawing the basic relational mechanisms that facilitate the desired behaviours and hinder the unwanted ones.

   This means working with policies "by design", which give shape to the conditions of possibility of behaviours that one wants to determine or modify. Designing such conditions means creating relational mechanisms that work not merely according to a logic of control and of possible sanctions, but above all according to a logic of reflexivity of self-reinforcement: virtuous circles such that the more they work, the better they work. For example, the widespread interest of citizens in the use of digital payments instead of cash can result, as a beneficial side effect, in greater tax control on the transactions themselves and therefore on tax evasion, a decrease of which could lead to a reduction in the tax burden, an improvement in the economy and greater incentive to use digital payments, and so on. It is therefore a question of technically designing virtuous circles that improve society and which are strengthened the more they are used.

20. Good politics pursues its aims, including its human project, through the promotion of *economic well-being*, freely enjoyed or sought by people, not through the exercise of *coercion*.

21. Good politics does not use *coercion* as a means but, classically, maintains its *monopoly on violence* to eradicate it altogether, or replace it with peaceful, equitable, sustainable, and productive *competition*.

22. Good politics is guided by good ideas in satisfying, reconciling, and prioritizing, within its human project, the reasonable hopes and legitimate interests of people and society, with regard to individual, social, and environmental prosperity.

23. Ideas are good when they provide politics with strategies that are *feasible* (achievability), *efficient* (cost), *effective* (result), *shareable* (consensus), and *desirable* (ethics) to take care of individual, social, and environmental prosperity.

24. Good ideas are generated by good *reflection* and are consolidated by good *practice*.

25. Reflection is good when it is *rational* in its reasoning, *informed* about facts, aware of its *fallibility*, *tolerant* of different opinions, and open to *constructive dialogue*.

26. Reflection takes place in the *public sphere*.

27. The public sphere is part of the *infosphere*.

28. A practice is good when it is *transparent* in the sense that is both accountable and auditable.

29. A good reflection is promoted by a good political debate.

30. A political debate is good when it is based on a good reflection and decides, in a satisfactory way, on the goodness of the available ideas, on their compatibility

and priorities, and on how to achieve them, creating a fair and open market of tolerant and just ideas.

31. Good ideas are not partisan but, because of their nature, they are *shareable* by more than one political program.

Knowing how to recognize and support good ideas, regardless of the source and the context that offers them, is essential in a political context that is increasingly "on demand" and "just in time" and less and less "always on", in which the management of the attention of the civil society must be based on the forward-looking interest in the proposing of good and relevant ideas, and not on alarmism, emergency, or recurrent crises.

32. Sharing good ideas regardless of line-ups or political programmes means privileging *ethics* to *ideology*.

33. Good ideas motivate politically (in a sort of political *psychagogy*[21]) by relying on three factors: *hope* (which can also be altruistic and public, and when negative can become *envy*), *interest* (which is usually only personal and private, and when frustrated can become *anger*), and (inclusive disjunction) *reasonableness* (from common sense to logic, from the correct use of facts to probabilistic reasoning).

34. *Hope* motivates more than *interest*.

There is no personal interest—including the fundamental one for one's own well-being or that of others, and for one's own survival or that of others—that cannot be overcome by hope, to the point that people can commit suicide because of their hopes. For this reason, fundamentalist or ideological terrorism, when it is driven by hope, cannot be fought or counteracted by appealing to interest.

35. *Interest* motivates more than *reasonableness*.

There is no reason, including mathematical certainty, which cannot be neglected, perverted, or underestimated for personal interest.

36. The *hubris* of reason consists in its faith in the cogency of its own epiphany.

In other words, reasonableness (the epiphany of reason) is not necessary and can be insufficient (is not cogent enough) to motivate politically. Reasonableness is reconcilable with hope and interest but motivates less than either. This follows from the previous points. It is why the most rooted greed, which is based on selfish interest, cannot be fought by appealing to reasonableness. In particular, social problems—above all, corruption, fundamentalism and intolerance, exploitation and violence—and environmental problems—above all global warming, biodiversity loss, pollution, and violence on animals—cannot be solved by leveraging only reasonableness as motivation.

37. Good politics is successful if it motivates above all on the basis of hope, then of interest, and finally of reasonableness.

---

[21] In ancient Greek philosophy and early Christian theology the term refers to "guiding the soul", e.g. through reflection and education about correct conduct and the obtainable virtues. Today, it refers to attempts to influence a person's behaviour, e.g. by suggesting desirable life goals.

A winning political campaign, from Berlusconi to Trump, from Brexit to the populist movements in Italy, devalues the present, that everyone has an interest in changing as always unsatisfactory, and overestimates the future, that everyone is hoping to be better. A losing campaign, from Hilary Clinton to the Remainers and the defeat of Renzi and his Partito Democratico (PD) in Italy, values the present as already satisfactory, often indicating how much better it is when compared to the past,[22] disappointing the hopes of all those who want it to be better; and evaluates a possible future as worse or risky if the alternative wins, thus frustrating the electorate's hopes, promising only a reasonable yet unattractive *more of the same* (another Clinton presidency, the usual European Union, another Renzi government), that is, a losing political message.

38. *Fear* is only an indirect motivational basis.

   This is because anyone who has no hope, or has no interest, or does not listen to any reason, cannot be motivated by fear. Fear works only if it frustrates or threatens hope, interest, or reasonableness.

39. *Punishment*, understood as an instrument for the management of fear and therefore of interest, is always ineffective if it generates desperation, understood as a total lack of hope.

40. *Public opinion* is born of the hopes, the interests and the reasonableness of the public that expresses it.

   Public opinion is rarely reasonable (it is not an expression of *nous*), it is often above all *emotional*, in terms of hopes and fears, and *instinctive*, in terms of interests (as an expression of *doxa*). Therefore, its formation is very rarely deliberative but above all psychological and hence rhetorical.

41. The *rhetoric of reason* is the best way to shape public opinion politically.

   Good ideas alone are never enough; they need to be explained and supported in a persuasive way.

42. Good ideas are *timely* (they work at the right time) not *timeless* (as if they worked any time), and therefore *dynamic* and always *updatable*.

   This is because the solutions they propose are *not immutable*, like the laws of nature, but *contingent*, like human history, and must evolve with the problems they face. The timeliness of good ideas is neither *relative*—as if it depended entirely on circumstances and always and only changed with them— nor *absolute*—as if it did not depend on circumstances at all, and never changed in relation to them. It is *relational*, because it depends in part on the circumstances and changes interactively with them, trying to improve them.

43. It is on good ideas, their priority and feasibility, that *consensus* must be created.

44. Consensus is the cooperative and contextual convergence of relations.

45. The two fundamental values that qualify political relations are *solidarity* and *trust*.

---

[22] See the list of U.S. presidential campaign slogans: https://en.wikipedia.org/wiki/List_of_U.S._presidential_campaign_slogans

46. Politics as a practice is the totality of *solidarity* and *fiduciary* (trust-based) relations that organize and guide a society.
47. *Solidarity* regulates needs in a society and is at the root of *green* (environmental and ecological) solutions.

     This is solidarity understood as the mutual care of relations with others, with the world, and with future generations. Without this solidarity there is only a free market but no fair prosperity.
48. *Trust* regulates actions in a society and is at the root of *blue* (digital) solutions.

     This is about trusting ourselves, each other, the future, human ingenuity and its products, and the potential goodness of their applications. Without this trust there is only management of political power and a market of people's views, but not also a good policy and a market of ideas.
49. Politics takes care of the relations that make up and connect things.

     Focusing on the primacy of relations rather than on the primacy of things—for example, on the primacy of the concept of "citizenship" rather than that of "citizen"—means that good politics must move from taking care of the good management of the *res publica* to taking care first of all of the nature and the healthy growth of the relational network that constitutes a society, its members, and its environment, that is, the *ratio publica*, as previously defined. The fabric of the *ratio publica* is the inter-spatiality of historical–cultural relations that give identity to a society and its members.
50. Criminal politics is a form of *mafia*.

     Mafia replaces politics in taking care of the relations that make up and connect things. This is why it is incompatible with the State and survives only by becoming government.
51. Politics, when it does not work, can only be repaired if its relational nature is repaired.

     Politics is malfunctioning when the two main relations of solidarity and trust do not work. It can only be repaired by repairing the two relations. This should be a reason for some comfort and moderate optimism, because it is easier to repair relations than the *relata*, that is, the things constituted and connected by the relations. For example, it is easier to repair the relation of trust between two political parties than "repair" the political parties themselves to make a relation of trust work.
52. Good politics is metaprojectual, that is, it supports the *individual human project*.

     Every individual is a path of self-realization, through which a person progressively becomes more and more himself or herself. This individual, open and autonomous construction (*poiesis*) of the self is a delicate process because every individual does not exist in their own right and alone, but comprises a knot of relationships, fragile, flexible, and easily influenced and damaged. Politics supports individual self-construction (*autopoiesis*), providing the conditions for its realization, especially in terms of tolerance, justice, peace, freedom, security, education, respect and recognition of others, and equal opportunities. Politics is malfunctioning when any of these conditions is not met.

53. Good politics support the *human social project*.

Every society is in constant tension, even if only implicitly, towards the realization of what it would and should be, that is, as a shared and shared human project, which is an open-ended work in progress. Politics is concerned with supporting and implementing the best possible human social project, in a critical and conscious way, that is compatible with the historical circumstances in which it arises, and the individual human projects of which it takes care.

54. A fundamental value promoted by good politics is *just tolerance*.

Starting from Locke, tolerance lies at the root of the modern political era, as a request to keep every individual and social human life always open to choice, change, and rethinking. Tolerance must be just, i.e. attentive to the negative effects of its excessive application. But justice itself must be tolerant of difference, of error, of the possibility of doing otherwise or better, of starting again, and should not rely on the excessive application of protocols and automatisms. Justice recognizes the logical superiority of tolerance when it assumes, as its own limit, the acceptance of unjust injustice rather than unjust punishment: better a criminal outside prison than an innocent in prison. Hegel was right (*pereat mundus ne fiat iustitia*) not Kant (*fiat iustitia, pereat mundus*).

55. The exercise of just tolerance promotes the care for *human fragility*.

56. Respect for human fragility should be a *universal right*.

Individuals are delicate informational organisms, open and adaptable to change, malleable by education and imitation, transformed by events, changed by circumstances, influenced by the flow of information and the informational environments in which they find themselves. The first duty of politics is to ensure that human fragility is always respected and never exploited no abused.

57. Politics does not *log out*.

Socio-political relations can be modified but not denied. So, the rhetoric of being inside or outside (for example of Europe) is made hollow by the fact that, in a global relational network (*cosmopolitanism*), one cannot be disconnected, but only connected, and this in a more or less correct and coherent way with the social human project pursued. Bad politics does not disconnect (log out) but badly connects (short-circuits) the social relations and interfaces that must facilitate and coordinate them. The impossibility of politics to log out is the new embodiment of the old-fashioned, Aristotelian idea of politics being always-on. The continuous political nature of everything that happens in a society (no log-out) should not be confused with the discontinuous political nature of social engagement (politics is now on-demand).

58. Politics is *cybernetics*.

In Plato, the kybernetes or "steersman" is the pilot of the ship, which navigates in the right direction, even against the current or unfavourable winds, and therefore sometimes indirectly and obliquely. Politics' main task is not to manage the *speed of change* (for example technological innovation), but to determine the goodness of the *direction of change*. It may or may not have a foot on the brake or the accelerator, but it must have hands on the steering wheel. The high speed with which a society proceeds in its transformations can be a good thing, if the direction chosen by politics is the right one.

59. Politics is *Markovian*.

Like a chess game, politics is constrained by the past, but it knows only the present, to be managed and negotiated (and in case criticised), and the future, to be designed and planned (and in case promised). This is so because voters have no memory. Whatever politics delivered in the past, whether a problem or a solution, is taken for granted. The only past that is present in the voters' minds is unrelated to history and is part of a story-telling. So those who shape the narrative of the political past control its impact.

60. Democratic politics is *binary*.

Democracy is usually defined in terms of the *shared values* (*semantics*) or *rules* (*syntax*) adopted by a society. In reality, semantics and syntax presuppose a previous *structural step* of separation between two elements: *sovereignty* (possession of political power) and *governance* (the exercise of political power). Without this binary structural condition, a democracy flattens out into a dictatorship, in which the majority (which owns and exercises political power) imposes its will on the minority, whose individual or collective human project is not protected.

61. The space of politics is part of the *infosphere*.

Today the *space of politics*—understood also as *public space* (see above) and as a *deliberative exercise*—is always *onlife*: partly online and partly offline, partly analogue and partly digital. And it is so also for those who are still excluded from the digital revolution (those on the wrong side of the *digital divide*), because their choices are conditioned, influenced, or determined by those who are included.

62. Good politics today must make *capitalism sustainable* and *fair*.

Capitalism is the best system known to date to produce wealth, but not to produce it in a *sustainable* way (in terms of *environmental impact*) and to distribute it *fairly* (in terms of *social equality*). Good politics rectifies robustly these two limits, while supporting *private property*, *project ownership*, *competition, innovation, investment*, and *profit*.

63. Good politics today must replace *consuming* the world with *fostering* it.

In the past, capitalism has been seen as an inseparable counterpart to *linear consumerism*: producing, using, consuming, and disposing of things. But now this link can and must be severed, in favour of a new coordination between capitalism and the economy of caring for the world (that is, *circular fostering*). Moving from a politics of things to a politics of relations, it is easier to start building a post-materialist and post-consumer society, which privileges a *circular economy* of services and experiences in a fair and sustainable way.

64. Good politics organizes and manages a *capital of citizenship*.

Every generation enjoys the work, the efforts, and the sacrifices made by all the countless past generations, because each generation is the heir of past humanity and in turn leaves its legacy to the next generation. Politics in the twenty-first century should adopt strategies to distribute and capitalise on the benefits of inherited wealth, guaranteeing to members of society not only equal opportunities but also a capital of citizenship to support individual projects.

65. The State is an *interface* that performs a function of relational support for the creative and fruitful strategies implemented by a society.

The State is not the point of arrival of the legal–political organization of a *polity*—which we have seen to be a political community, that is, the political ordering of a society—but the relational meeting point—that is, a *dynamic interface*, that can be realised in a variety of ways—between polities, that is, between a society that organizes itself through it, and the other societies, organized like other States, in the rest of the world. Citizens interact politically among themselves and with the world through the interface-State, to which they belong, and the various interfaces within the State (e.g. at the regional or city level). Different dynamic interfaces allow this interaction and communication, which does not require a single model at all—think of the various models of State organization, for example federations, presidential republics, constitutional monarchies, and so on. In the digital age and globalization, it may seem that the State no longer has a key function, and that the alternative is either a more rooted localization and corresponding micro-nationalisms—see the many phenomena of independence in various European States, from Spain to Great Britain, from Germany to Italy—or a multinational globalization consonant with markets, large companies, and intergovernmental institutions. In reality, the greater the globalization, the more necessary is the State, understood as an interface of communication, interaction, and coordination between local and global realities. The crisis of the modern State is not a crisis of "necessity" but of "sufficiency": the State is increasingly necessary, but also increasingly insufficient, to take care of the *ratio publica*. It is joined by many other equally necessary agents: supranational organizations, international institutions, and multinational companies.

66. A State is good when it implements good politics.

67. Good politics enhances the State as an interface.

This is because the State can: invest in useful infrastructures; cover the risk of long-term investments; decouple the success of creative and productive strategies from formal (e.g., bureaucratic) or substantive (e.g., lack of credit, public debt) obstacles; stimulate the development of socially acceptable or preferable strategies, and discourage the development of unacceptable ones; rectify the limits of the markets when necessary; delegate the success of economic strategies, by outsourcing and controlling, whenever possible, any creative and productive activity that does not require State management, in favour of private initiatives; coordinate efforts at national and international level; and ensure that doing the right thing is not penalizing (equal conditions, level playing field). For example, digital technologies are disrupting the labour market. It is clear that, due to new forms of automation and information management, many functions—and therefore many related skills that today are part of a job—will soon no longer exist or will be significantly transformed, while new ones are not yet imaginable. In this context, the precariousness of present solutions and the uncertainty about future solutions will grow. A defence strategy, which sees resistance to digitalisation (think of the proposal for a tax on robots) and increased welfare as the only solutions, is not a winner. We need a strategy of attack, with more mobility, agility, and flexibility in the labour market but also, and for this very reason, a robust protection network. Because mobility, risk-taking, re-skilling, and entrepreneurship are sustained and facilitated by reducing risks and the costs of failure. It is never the poor who can afford to take risks and fail. The digital revolution can become a great engine for the generation of work opportunities if any failures in its exploitation are seen as normal, not penalizing, and are mitigated in their negative effects by a robust socio-economic support culture. A culture of "little gain, but sure gain" will not bring any country to the top of the digital economy, but will instead make it slip into a rear-guard position. Instead, a strategy for a digital economy of "much and uncertain gain, but certain protection" is needed. In this case too, I claim no originality. As Matthew 13.12 states, "For whoever has, to him more will be given, and he will have abundance; but whoever does not have, even what he has will be taken away from him".

68. Good politics is *multiagent*.

The State has the convening power and the duty to coordinate (infraethics) other agents to take care of the *ratio publica*. Above all, the State should call all the stakeholders, including the corporate world, to share the responsibility, in a visible (transparency) and responsible (auditable accountability) way, of making policies together, in a multiagent pact guaranteed and managed by the State itself. This is also true at the supranational level, where the European Union, for example, has the strength and the duty to coordinate other States and stakeholders to take care of the European *ratio publica*.

69. Good economic policy is an economy of *onlife experience*.

The time available and its quality are the most important (finite, non-transferable, and non-renewable) resource for every individual. Therefore, the prosperity of individuals, their societies, and their environments is also assessed on the basis of the management and enrichment of their individual and social time. The modern era is widely interpreted as the period during which humanity has managed to "heal" more and more time—especially thanks to the improvement of living standards, scientific research, and national health systems—and to "free" more and more time—especially thanks to the various phases of industrialization and technological development, to trade, and to socio-political conditions. We live longer and better than any other past generation; and we live with much more time and income at our disposal. This is why, today, an innovative economy of growth should focus on the management and enrichment not so much of working time, but of healthy or healed time—that is, the time spent without suffering and illness—and of leisure or liberated time—that is, the *disposable time* (in analogy to disposable income), which is *available* and onlife, not bound by work commitments, and *usable*, that is allocable to activities of choice. In a world in which healthy time and free time will increasingly expand, the corresponding economic activities linked to their intelligent management and their fruitful use will be increasingly crucial. The future of advanced economies is not in the consumption of things but in the enjoyment of experiences.

70. The solutions of good politics are *green* and *blue*.

The marriage between nature (*phusis*) and technology (*techne*) is vital for the prosperity of the planet, its inhabitants, and therefore every society. Today, the solutions found by good politics, in order to design and pursue the human project for a mature information society, must be both green (environmental and cultural economy and policy), and blue (digital economy and information policy). Environmental, artificial, cultural and digital environments must be fostered to ensure that they coexist in symbiotic relationships of mutual benefit. Not only must they be protected, but they must also be valued as resources for individual and social well-being, and not wasted. And they must be taken care of in a holistic way. This also means that the mentality of the exclusive protection and care of environmental and cultural assets—the environment and culture as a burden and cost for society, education included—should be transformed into an economic strategy of promotion and utilisation, seeing the environment and culture as precious capital to be put to use, for the benefit of the whole society that expresses it, and dependent on digital technologies.

12.7   Conclusion

When we talk about the digital revolution it is natural to ask ourselves what the next radical transformation will be. Human history certainly does not end here, and there will be other extraordinary changes that we cannot even imagine. These are real unknown unknowns. Just think of what we would have answered, say, in 1920, if someone had asked us to predict the future in 2020. It is simply unimaginable what the world will be like in 2120. That said, the right perspective is that digital technologies will certainly bring other incredible innovations, but the transformation from an entirely analogical world to one that is also (and in some places, perhaps above all) digital has already happened. More will happen, but not this. Our questioning is a bit like wondering what else to expect after arriving on a new continent. We have "landed" on the digital, and we have mapped only the coasts (to continue the analogy), but the historical step has been taken. A small one for this generation, but a giant leap for future ones. So now the most important revolutionary challenge is understanding what to do with this new continent, all to be built. In other words, the new real challenge is not digital innovation but the *governance* of the digital. Digital governance is currently delegated (or abrogated) to the corporate world—primarily American—which follows a logic of profit-seeking and implements an entrepreneurial culture. This is fine in itself, but it is also an unsatisfactory solution as a whole because it risks ending up as a colonising monopoly—while missing the immense, counterbalancing contribution from (and to) the rest of society. However, to support and complement a necessary but insufficient corporate governance of the digital, we need above all good political strategies and courage in making the right social choices. In other words, there is a great need for good politics.

Acknowledgements This text is a revised and updated version of a pamphlet that I published in March 2018 in Italian with the title "Il Verde e il Blu—Idee ingenue per migliorare la politica in una società matura dell'informazione" (Rome: Formiche). The project was begun in 2016, after Brexit and before the election of Trump and the huge success of two populist parties (the Northern League and the Five Star Movement) in the Italian general election of 2018. The idea of this text took its initial form during some meetings with Alessandro Beulcke, with whom I conceived the project, discussed its contents, and shared the enthusiasm for the value that an informed, rational, intelligent, and open dialogue may have to improve politics. Without Alessandro, I would probably never have had the courage or the confidence to embark on this project. So, I consider him co-responsible, but only of the good bits, the rest is my fault. The support by Allea, Alessandro's communication company, was crucial for the logistics side.

Besides Alessandro, several people helped me to improve the ideas contained in this chapter in a profound way. They did it with a lot of patience, kindness, intelligence, and a remarkable investment of time. I really do not know how to thank them. For privacy reasons I will not name all of them. But I will nominate a group (which I have sometimes called G 18:3, with reference to Matthew 18:3) which has contributed to various versions of this text in a critical way, in addition to Alessandro himself: Alessandro Aleotti, Monica Beltrametti, Barbara Carfagna, Luca De Biase, Adrio Maria de Carolis, Massimo Durante, Ugo Pagallo, and Sergio Scalpelli. In addition, I would also like to thank Fabrizio Floridi, Kia Nobre, Stefano Quintarelli, and Mariarosaria Taddeo, who have read many versions and provided more suggestions on how to improve them than I can remember. Finally, David Sutcliffe commented and edited the English version with so much acumen that it became a new edition. To all these people, named or not, goes my deep gratitude. If the things I wrote in this text are not entirely wrong, it is thanks to our discussions and the feedback I have received from them. None of the people mentioned is responsible for the errors that remain, which are only mine, and I mean it. And above all, none of the people who helped me have approved or subscribed to the ideas presented here, and for which I take full responsibility.

References

Cassirer, Ernst. 1923. *Substance and function, and Einstein's theory of relativity.* Chicago: Open Court.

Floridi, Luciano. 2008. Artificial intelligence's new frontier: Artificial companions and the fourth revolution. *Metaphilosophy* 39 (4/5): 651–655.

———. 2009. The information society and its philosophy. *The Information Society* 25 (3): 153–158.

———. 2011. A defence of constructionism: Philosophy as conceptual engineering. *Metaphilosophy* 42 (3): 282–304.

———. 2014a. *The Fourth Revolution – How the infosphere is reshaping human reality.* Oxford: Oxford University Press.

———. 2014b. *The onlife manifesto – Being human in a hyperconnected era.* New York: Springer.

———. 2015. Toleration and the design of norms. *Science and Engineering Ethics* 21 (5): 1095–1123.

————. 2016. Tolerant paternalism: Pro-ethical design as a resolution of the dilemma of toleration. *Science and Engineering Ethics* 22 (6): 1669–1688. https://doi.org/10.1007/s11948-015-9733-2.

————. 2017. The logic of design as a conceptual logic of information. *Minds and Machines* 27 (3): 495–519.

Halmos, Paul R. 2017. *Naive Set Theory*. Mineola: Dover Publications, Inc.

Index

© Springer Nature Switzerland AG 2019
C. Öhman, D. Watson (eds.), *The 2018 Yearbook of the Digital Ethics Lab*,
Digital Ethics Lab Yearbook, https://doi.org/10.1007/978-3-030-17152-0